Rad

– safer solutions for cell phones
and other wireless technologies

by Kerry Crofton, PhD

Revised and updated 2nd edition

The Mom's Choice Awards® (MCA) has recognized *Radiation Rescue* with Gold – its highest award. MCA is known for establishing the benchmark of excellence in family-friendly media, products and services.

Radiation Rescue
– safer solutions for cell phones
and other wireless technologies

www.radiationrescue.org

Copyright © 2010 by Kerry Crofton, PhD
Revised and updated 2nd edition

All rights reserved. No part of this book may be used or reproduced by any means, graphic, electronic, or mechanical, including photocopying, recording, taping or by any information storage retrieval system without the written permission of the publisher except in the case of brief quotations embodied in critical articles and reviews.

The views expressed in this work are solely those of the author and do not necessarily reflect the views of the publisher, and the publisher hereby disclaims any responsibility for them.

Global WellBeing Books

ISBN 978-0986473500

Printed in the United States of America

Cell phones and other wireless communication technologies are a real danger, especially because of their unchecked growth, and the clear evidence of harm. Our young people, in particular, are at risk.

Professor Olle Johansson, Karolinska Institute, Stockholm, Sweden

Here is the truth that may have been hidden from us

Professor Olle Johansson and other many other experts reveal that these devices have not been properly tested for health safety, and that our government standards are way out of line with the science and are not protecting us.

You will find this book interesting and useful if:

- you have heard reports of cell phone health hazards and want to know how to protect yourself and your family;

- you want to take steps to reduce your risks but need some sound scientific evidence for the doubters;

- you want to know what the experts say about the safer use of cell phones, headsets, cordless phones, Internet access, game stations and other electronics;

- you – and/or someone you care about – spend a lot of time in front of a computer, in a wireless environment, on your mobile phone, or traveling in high-exposure environments like aircraft – even without feeling any ill effects;

- you are experiencing "unexplained" sleep problems, headaches, pain in your eyes, heat/pain on the side of your head/ear, dizziness, numbness and/or tingling in your hands, or difficulty focusing;

- there's a young child, newborn, or one about to be born, in your family;
- you want the facts on ultrasound, X-rays, mercury fillings and vaccines;
- any of your children have learning difficulties, ADD, ADHD or symptoms on the Asperger's and Autism Spectrum;
- you, or any of your family and friends, are suffering from any of the following conditions addressed in this book by our distinguished scientists and medical experts:

- allergies
- ALS
- Alzheimer's
- ADD
- ADHD
- Asperger's Syndrome
- Autism Spectrum Disorders
- cancer
- cardiac conditions - arrhythmia, tachycardia, hypertension, angina and others
- cataracts
- Chronic Fatigue Syndrome
- depression
- dizziness
- electro-sensitivity
- headaches
- infertility
- insomnia
- immune suppression
- migraines
- Multiple Chemical Sensitivity
- Tinnitus
- Parkinson's

If none of these issues catch your attention, you may pause at the information on how to boost your immune system and stay well as you age in this hyper-tech world. You will also discover why the concern is not just about your own health but the wellbeing of all life forms on this planet.

Dedication

May I offer this dedication to you,
the person reading this book,
for considering this wireless wake-up call.
May this book bring benefit to you and your loved ones.

Gratitude

I salute the courage and dedication of the women and men who have been researching this issue for so many years. I am deeply grateful to them, and to the many other experts who have contributed generously to this book.

My husband and children have been wonderfully patient and supportive. Many thanks to them, and to our youngest daughter for the book's title. My never-ending appreciation also goes to Sherry Lepage for her inspired editing skills, and to Dr. Heather McKinney for her scientific expertise.

Reviews of the first edition

Generation Rescue, an international movement of scientists and physicians researching the causes and treatments for autism, ADHD and chronic illness, offers this:

> Our community is very concerned about the effects of electro-magnetic radiation (EMR) on the body, how it deregulates the ability to manage toxins and infections, and how it seems to interfere with blood flow in the brain. We feel that Dr. Crofton's approach to reducing EMR exposure should be looked at by all families looking to raise a healthy child.

This once-sleepless mom heard our wireless wake-up call – loud and clear – and took action right away:

> I found the information so compelling that I decided to make some immediate changes. I have been struggling with insomnia for the past three years. The first night after clearing my room of electronics, I slept like a baby. Likewise for the next night. While I am willing to consider a certain amount of placebo effect, I have now been sleeping well and deeply for a few weeks. This is one of those life-changing books.

This dad became so electro-sensitive that he sought not only to regain his own health, but also to prevent a similar situation with his young family:

> I saw that we were making choices that involved sitting in front of a computer, TV or video screen, and were using our cell phones and computers with wireless Internet much of the day. As I unplugged more myself, I was sleeping much better and had more energy.

Another parent noted:

> *Radiation Rescue* has made me more aware of the digital addiction that my family and I have nurtured. This book has had a profound and beneficial effect on our family.

An expert in the field commented:

> I found this an easy read, which is rare for this technical topic. The presentation is user-friendly. One is not overwhelmed in the learning process, but assisted by the relaxed and friendly style of the author.

A mother turned local activist wrote:

> Thank you so much on behalf of all mothers out there who have fought for the health and safety of their children. I hope everyone reads about the baby monitors.

A professional couple working in the computer industry reported:

> We had heard about cell phone health concerns but had no idea there was so much evidence. Your room by room audit was a wake-up call in itself. Like many people we had a cordless phone and an alarm clock/wireless unit right beside our bed, a fluorescent reading lamp, and cell phones charging on the bureau – in our wi-fi home. No wonder we couldn't sleep. Making the changes your experts suggest was not difficult. (We've now got the recommended headsets.) Your book gives us hope.

An advocate in the UK recommends this book:

> *Radiation Rescue* is a useful reference for anyone living in this wireless age and concerned about any element of their health, or lack of health. There has been very positive feedback from people in the UK concerned about this issue.

A note on this revised edition:

A few readers, who were already suffering health effects, were upset with my suggestion that we could use this technology more safely, and for not communicating enough urgency. We do need to act now, as wireless mobile devices and communication networks are becoming increasingly powerful, and widespread. May I also say we need to be realistic, listen to this team of authorities and heed their warnings and recommendations. We can face this challenge with skilful action, not fear.

I have updated this wake-up call with new evidence, clarified a few technical details, and invited more top experts including a cardiologist who explains some 'unexplained' cardiac symptoms, and four more highly qualified technical experts.

And there is good news to report!

US senators, who held a hearing on cell phone safety in September 2009, promise to pursue this issue and the French Senate has just passed a motion banning the use of mobile phones in all primary and middle schools.

And it's not just legislators speaking out: independent scientists are being interviewed in mainstream media; concerned moms are forming advocacy groups; feisty citizens are challenging the installation of cell tower antennas and power lines; wireless radiation is being removed from some schools and public spaces.

We have created this situation. Working together we can turn it around.

Please keep this hope in your heart as you read on …

These children are asking us the key question.

Having a digital wireless device in your home, office, or school is like having a mini base station (cell tower) indoors with you. A wireless unit (including most cordless phone bases) is emitting radiation virtually constantly, all the time it's switched on.

Your body is more sensitive than any wireless device, so if your laptop, cordless phone or mobile phone can pick up a signal, it means that your body and your brain cells are receiving those waves, as well.

The nature of this wave is different from radio or TV; it is a pulsed wave, like a saw tooth wave. And many hundreds of studies strongly support the view that this type of radiation is responsible for a whole range of adverse health effects, including: suppressing the body's own immune system - making us more vulnerable to cancer and other illnesses - and interference with normal cell functions of the type that govern every detail of your body's behaviour.

These effects have been shown to occur at levels significantly below what governments consider safe.

Grahame Blackwell, PhD – Chartered Engineer, former leader of a cell phone R&D team, computer expert and advisor to WiredChild (www.wiredchild.org)

Earth Friendly

This book is about dealing effectively with global electro-pollution. It's also essential to be aware of other environmental footprints, as you know. The author, her husband, who used to run a reforestation business which has planted millions of trees, and their children have requested that this book be printed on 100% recycled paper or, at least, a combination of recycled paper and wood pulp not harvested from old-growth or endangered forests. In consultation with environmental groups, they have also made a commitment to have trees planted to offset the book's environmental impact.

Disclaimer

Please note this material is not meant to be a substitute for the advice provided by your own physician, or other health professional, and should not be used for self-diagnosis, or for treatment without medical supervision.

Neither I, nor the experts offering their research and/or recommendations, assume responsibility for how you use this information, or for any circumstances arising out of the use, misuse, interpretation or application of any information supplied in this book.

I am not an expert in the health effects of EMR, and am not qualified to give an expert opinion.

The opinions expressed by me are my personal observations, and not intended to malign or defame any industry, agency, individual or product.

About the Author

Kerry Crofton is a health educator and concerned parent. She is the director of WellBeing International and founder of Radiation Rescue, an educational organization dedicated to giving you the information you need to safeguard yourself and your loved ones, in this digital age.

Kerry has a doctorate in health psychology, and her thirty years of work includes: a clinical practice, developing and delivering stress management and wellness programs for air traffic controllers, commercial and fighter pilots, nurses, teachers, parents, and others in high-stress occupations. She directed a hospital-based cardiac program, and produced a television program on heart health. For several years, she wrote a newspaper column.

Some years ago, she wrote *The Healthy Type A - Good News For Go-Getters,* Macmillan, 1998 about her experience working with heart patients.

A lifetime of exposure

Dr. Andrew Weil is following this issue and offered us this quote:

> Since cell phones show no signs of going away - indeed, most American children today face a lifetime of exposure - it's vital to focus closely on the most recently published studies, the ones that show the effects of longer-term exposure.
>
> You may have heard about a recent report from the Environmental Working Group (EWG), a Washington, D.C.- based nonprofit organization that advocates for health-protective policies.
>
> The EWG reported on key studies (including some published from 2007 to 2009) that link radiation from long-term cell phone use with increased risks of brain and salivary gland tumors, migraines and vertigo, as well as behavior problems in children, including hyperactivity.

Andrew Weil, MD, Founder and director of the Arizona Center for Integrative Medicine and author of *Healthy Aging: A Lifelong Guide to Your Well-Being* (Knopf, 2005)

Table of Contents

Overview   1
    4 Steps To Protect Our Families   13
    My Wireless Wake-up Call   16
    Contributing Researchers, Clinicians and Technical Experts   25

Step 1. Know the Evidence   29
    Microwaving this Blue Planet   30
    A Bit Of History   34
    Agencies Passing the Buck?   44
    Biological Effects - Evidence of Damage   50
    Safe Levels? A Difference of Opinion   78
    There's reason for hope   95

Step 2. Know Your Risks   109
    The Radiation Rescue Questionnaire©   111
    The Electro-magnetic Spectrum   123

Step 3. Seek Safer Solutions   131
    The Sources and Recommendations   135

Step 4. Enjoy Your Family Action Plan   237
    Room by Room – Audit & Recommendations   238
    The River of Health – pristine or polluted?   255
    EHS Symptoms and Related Conditions   262
    The Low-Tech Family Challenge   344

Hope For The Future   359

Resources   381
    Research, Advocacy and Electro-Sensitivity Support Groups   381
    Autism Related Issues   387
    Children & Nature   387
    Medical Clinics/Professional Practices   388
    Professional Services   392
    Reference Material   402

Index   423

Robert Becker, MD was a surgeon and author of *The Body Electric* (with Gary Selden, Harper 1985) and *Cross Currents – The Perils of Electropollution, The Promise of Electromedicine* (Tarcher Penguin 1990). Dr. Becker cautioned us:

> Our bodies and brains generate electro-magnetic fields within us and around us. We live in the Earth's natural magnetic field, and from the beginning we have been dependent on this environment. However, we have now created a vast global network of man-made electro-magnetic fields – the greatest polluting element in the earth's environment. The human body electric and the Earth's body electric have been damaged.

Stephen Sinatra, MD, is a cardiologist who is also very concerned:

> We have seen clinical evidence that electro-pollution affects the normal functioning of the electrically sensitive heart, including the rate and rhythm, and other systems of the body. All forms of pollution impair our health, and the electro-pollution arising from the proliferation of wireless technology has greatly added to this burden.

In 1962, in her groundbreaking exposé, *Silent Spring*, Rachel Carson proved brilliantly that our human health is inextricably linked with our environment. Carson advised us:

> Technology is often so fundamentally at odds with nature, and the wellbeing of all living things, that it must be carefully monitored.

4 Steps To Protect Our Families

Step 1 Know The Evidence – If you are already well versed in the evidence, the scientists' concerns and the misinformation, jump ahead to the 2nd section. I encourage you to come back to the first step, at some point, so you'll be better equipped to keep your family safe, and to educate others.

If you find yourself becoming overwhelmed in Step 1, please skip ahead to assess your risk (#2), or skip to the devices and solutions section (#3), rather than abandoning this altogether.

Step 2 Know Your Risks – Here you find a detailed questionnaire to help you see where you're being exposed to electro-magnetic fields (EMF). Some people refer to this as electro-magnetic radiation, or EMR; this is the term we'll use. We include, in simple terms, what you need to know about EMR.

Step 3 Seek Safer Solutions – This step offers information on everything from the lowest radiation dental X-rays, to what you need to know about tucking your child into the back seat of a hybrid car. Details here include the kinds of cell phone headsets to avoid – and the one that is recommended, the safer alternative to cordless phones, what kind of cell phone to buy – and the best way to use it. There's also key information about Internet access. You'll discover practical advice about many other devices, and ways to reduce your radiation exposure and related health risks. Everyone in your family will benefit.

Step 4 Enjoy Your Family Action Plan – How to create a safer environment for you and your family, room-by-room. Our contributing physicians also offer what you need to know about electro-sensitivity, autism, immune health, dental health and other key factors in your family's wellbeing.

There's advice on preventing or healing from EMR exposure, and our contributing experts reveal how they protect their own families. We also look at effective ways to deal with techno-stress.

An author on parent – teen communication gives us tips on how to get our young people onboard. We look at the Low Tech Family Challenge: to unplug and reconnect. This is for families wired on wireless, who take on reconnecting with each other, and the natural world around us. You'll hear from a few families who have made this shift from virtual to real, from cyber isolation to rediscovering family fun.

You may find that you actually enjoy this challenge, and creating your own Family Action Plan.

Hope For The Future – The conclusion brings it all together with inspiration from the families following this challenge, and from dedicated advocates and courageous political leaders making a difference in their communities. Not just to make this a better world, but to turn back the tide of electro-pollution for our children's future, for the wellbeing of the natural world, and for all living things in our ecosystem.

Advocates and researchers from around the world encourage us with their hopes for the future – a challenge for some who have been lone voices in the wilderness. When you and I, with our families, in our own homes, take on this Radiation Rescue we can thrive, not just survive, in our digital age.

We don't feel this radiation and we think it's not doing anything, but it's a very potent biological agent.

Martin Blank, PhD, Columbia University

My Wireless Wake-up Call

Oh no, you may be thinking, not another dire health warning. Like most people, I often feel fed up with alarming medical news and conflicting studies (one recent report implicated wine as a risk for breast cancer, and the next day another praised wine as protection from Alzheimer's Disease). What a choice!

Do you remember when it was a pleasure to sit back and read the newspaper – perhaps with a glass of wine? Most have had enough of the 'sky-is-falling' alarms. Yes, it's tempting to hang up on this wake-up call.

But please don't! There are effective solutions in this book.

First, we need to look at how we have become cocooned in convenience, in radiation hot spots, as it turns out. Realizing this hit me hard. I have been a health educator for thirty years yet have never faced an issue as challenging, and personally unsettling, as this one.

Briefly, here's the problem, as I understand it

As the physicians Dr. Robert Becker and Dr. Stephen Sinatra reminded us, we humans are magnificently designed bundles of energy living within the Earth's varied electro-magnetic fields. The strong forces of the Earth, Sun and Moon affect us, and we have adapted to these. Yes, we can be killed by a bolt of lightning, or develop cancer from too much sun, but overall we have evolved to deal with this natural radiation. However, we're now in a massive soup of human-made radiation, on top of limitless numbers of new chemical toxins – two unnatural exposures.

All the parents I know are concerned about raising children in an environment awash in chemical hazards. Let's demand the truth about the products we buy, and a shift away from the profit over people mindset.

We were justly horrified to discover there was lead in toys and baby bibs, toxic Bisphenol A in plastic baby bottles, carcinogens in baby bath products, an array of chemicals in plastics in just about everything from plastic wrap to shower curtains, and melamine in many pet foods, chocolate bars – and baby formulas! And now we hear there are traces of mercury in some high fructose corn syrup, an ingredient in many processed foods and sugary drinks.

Mercury, one of the deadliest heavy metals, is also often found in tuna from our polluted oceans, in most vaccines, and 'silver' dental fillings. My dentist once remarked, "If mercury is such a hazardous material it requires a strict protocol for handling, then how could your patient's mouth be a safe place to put it?" Mercury in dental fillings has been banned in Sweden and Norway. More on this issue when we look at dental health and at autism.

There's also a seemingly endless list of carcinogens added to color, flavor and preserve our food; toxic chemicals added to our cosmetics, cleaning products, detergents and dryer sheets. Purify and freshen the air with chemically-scented candles, and plug-in or spray room deodorizers? What are we thinking?

Whether we're buying laundry soap, children's toys, or cell phones, are we assuming that public officials and manufacturers ensure products are rigorously tested for health effects before they are allowed on the market?

I see now that I was relying on the fox for the safety of the hen house. I hadn't realized that it evidently takes government regulators a few decades *after* a marketplace exposure before issuing public health alerts.

Remember asbestos? Tobacco? Lead? (And then, there's the story that the ancient Romans did themselves in with lead-lined wine flasks.)

As you know, environmentalists have, for decades, been raising the alarm about the toxins we eat, drink and breathe. Remember the ground-breaking courage of Rachel Carson, author of *The Silent Spring,* who took on the chemical companies about DDT and other poisonous pesticides – and won. Carson, the mother of the environmental movement, warned us that the chemicals in our water, soil and air burden our systems and can lead to illness. Toxic sludge and smog are, at least, hazards we can usually see.

Imagine how horrified Rachel Carson would be about the environmental impact of electro-pollution, and its connection to the disappearance of the bees. And the birds. Her prediction of a silent spring now seems even more of a wake-up call.

Throughout this book, authorities explain how the birds, bees and people of all ages are being affected by EMR. Some of this may surprise you. Haven't we had radio and TV waves swirling around our heads, and microwave ovens and portable phones in our homes, for decades?

And aren't we living quite happily with wireless technology without any apparent harm?

Even if we realize the risks, we may not know how to protect ourselves, or our children, because the problem has been swept under the rug for so long. Not even the mainstream medical community seems aware of this issue; few recognize the related symptoms and conditions – let alone know how to treat them.

It has taken me some time, and a lot of researching, to discover how to reduce our exposure without unplugging from the power grid and moving to a cabin in the woods. I'll admit, I was resistant to hearing the disturbing evidence.

The disbelief. The worry. The guilt. The overload of depressing information. I didn't want this bad news….but now that I know what to do, I feel strengthened. And I want the same for you.

Offering wake-up calls, however, is a challenge.

Have you heard this: Well, everything gives you cancer, so what's the point of worrying? There's nothing we can do so why bother? Feeling disenfranchised, overwhelmed and powerless are obstacles that can be overcome, as you know.

I am hopeful that as this message is being broadcast more clearly, more people will hear not only that there is a problem, but most importantly, that there are solutions.

Please remember throughout this book — and when trying to wake up others — that we are, at this time, dealing with areas of scientific uncertainty.

Keep this in mind if you're talking with scientifically-minded people who may reject claims of harm that do not meet these absolute criteria for the burden of proof.

As far as I know to date, there are no long-term, controlled, double-blind studies that absolutely prove the link between electro-magnetic radiation and diseases like cancer, for example.

As you know, we are dealing with a new technological phenomenon that has blitzed the globe like wildfire. And as the scientific method requires a control group of people who haven't been exposed to EMR, where on earth could we find them now?

Dr. Heather McKinney, a researcher who cares deeply about the health effects of EMR, offers a few words of wisdom:

> By the time the desired 'scientific proof' from long-term studies makes it to publication, technology — and levels of exposure — will have already moved beyond that which was originally studied, making the results arguably, to some, obsolete.

> And most researchers often do not have time to spread the word of their findings. They must continue with their research and publish in academic journals.
>
> Unfortunately, public awareness and education are often missed in this cycle. Thank you for gathering material from the leading scientists and clinicians, and for being such a good message carrier.

In the spring of 2006, I was concerned about the high voltage power lines close to our youngest child's school. Then I discovered that the school had wireless Internet access and a cell tower nearby that were also putting the children at risk.

Investigating further, I found out that many things in my own environment were adding to the problem. Like most people, I carried a cell phone.

I seldom used it though – instinctively it seemed that anything that could push a powerful signal through a cement wall might not be good to press against my head too often. Or my children's. (You may rethink whether arming your children or teens with cell phones is really a protective safety precaution.)

My office contained the usual must-have technology and electronic gadgets: the latest cordless phones, a computer – with wireless Internet, of course. So convenient. Being what is called an early adopter I brought the newest conveniences on board. And I took pride in my healthy home.

Then I had a professional electro-magnetic radiation assessment done.

Talk about an electric shock! Our youngest child's canopy bed – so beautifully decorated in pale peach and cream, with luminous stars sprinkled on the blue sky ceiling – was a hazardous hot-spot because of the radiation-reflecting metal frame. I gasped when the radiation detector screamed an alert. Who knew?

I whipped that metal frame out of there before we moved on to the next room. I couldn't even bring myself to give it to anyone and thrust it into the metal recycling bin.

I was also shocked when the cordless portable phone – the high-powered 5.8 GHz model, of course – sent the detector off the scale, almost as high as the microwave oven.

Doesn't everyone have at least one cordless phone? Haven't we lived happily with this convenience for years?

I read recently that a UK physician, the Harley Street practitioner Dr. David Dowson, urged caution with this surprising statement:

> Having a cordless phone is like having a mobile mast
> (a cell tower antenna) in your house.

I have since replaced all our cordless phones with safer corded landlines – more details on all of this in Step 3.

Testing a friend's microwave oven was also a jolt! Backing up from it, to see how far the radiation field extended, I crossed her kitchen floor and almost fell down the stairs leading into the living room. The radiation detector was still blaring at 25 feet. She had known that experts say microwaves can carry health risks, so she didn't use it very often, and always stood back when it was on – but not 25 feet away!

And, I heard recently that if you microwave a cup of water for three minutes, the water can set off the radiation detector – after you remove it from the oven!

Another shocker: being green at heart we had been replacing regular light bulbs with compact fluorescents. Guess what? Yes, they are energy-efficient but can emit 'dirty electricity'; more details in Step 3. My husband gasped, "Do you know how much those cost, and how much energy they save?"

I decided to wait until another time to give him the news about the electric hybrid car he was saving for. (You got it: sitting so close to the powerful unshielded electric battery …)

Then I learned how the radiation from the wireless router I had installed, so we could conveniently access the Internet all over the house, was also an unexpected problem. I had no idea.

As a mother, of course I am used to feeling guilty, but I truly shuddered when I realized how I had unknowingly put in harm's way my beloved family, and myself – and our new puppy snoozing near the comforting electric heater. I did wonder, however, why she usually moved farther away from it.

Quite honestly, at first I found it a lot to take in. So I understand if you are feeling overwhelmed by all of this. Like me, you may go through flashes (just what we need – more hot flashes!) of disbelief, denial, outrage, anger or guilt.

In a short time, however, I did go from being duped to determined! And resolved to use my skills as a health educator and coach to let other parents know this information.

There are other issues here, as well: from how our children can become physiologically addicted to gaming, to the emerging concerns about wireless Internet in aircraft – the health, and cognitive functions, of the pilot, crew and passengers.

Yes, aviation safety concerns! One of our contributors, Dr. Hans Scheiner, certainly caught my attention with this, "Some symptoms caused by electro-magnetic radiation like headaches, drowsiness, vertigo, nausea, loss of hearing and vision, and a lack of concentration are also a concern in transportation safety, including aviation."

This is another good reason to take this issue seriously – without bogging down in fear or denial – and take action to reduce our exposure. Dr. Scheiner, and more than two dozen other experts, show us how.

My goal is to offer you what you need to know to transform your family's environment into a low EMR sanctuary and to keep your family safe.

I speak to you as a mother. I am not an expert in EMR health effects. However, I am in communication with some of the world authorities and will call upon them to keep us up-to-date with accurate information. We need the truth about the electronics we buy, and the technology that surrounds us. These bulletins will be posted on our website.

With my personal commitment to help you as much as I can,

Kerry Crofton   www.radiationrescue.org

Contributing Researchers, Clinicians and Technical Experts

I've consulted these respected scientists, physicians, environmental health experts, and EMR technicians for their research and recommendations:

Martin Blank, PhD – Associate Professor of Physiology and Cellular Biophysics at Columbia University;

Carolyn Dean, MD, ND – physician and naturopathic doctor specializing in complementary medicine and environmental health;

Professor Devra Davis, PhD – founder of Environmental Health Trust and award-winning epidemiologist;

Leslie S. Feinberg, DC – chiropractor and developer of the energy medicine treatment NeuroModulation Technique;

Larry Gust – electrical engineer and Institute for Bau-Biologie® (IBE) certified Building Biologist;

Katharina Gustavs – Building Biology environmental consultant specializing in electro-magnetic field testing and environmental and occupational health;

Stan Hartman – environmental health consultant with Radsafe in Boulder, CO specializing in EMF/EMR issues;

Magda Havas, PhD – Associate Professor of Environmental and Resource Studies at Trent University;

Carrie Hyman, LAc, OMD – licensed acupuncturist and Doctor of Chinese Medicine who was an associate professor at the California Acupuncture College;

Olle Johansson, PhD – Associate Professor at the Department of Neuroscience, Karolinska Institute and Professor at The Royal Institute of Technology, Stockholm, Sweden;

Vini Khurana, MD – Neurosurgeon at the Canberra Hospital, and Associate Professor of Neurosurgery at Australian National University Medical School;

Dietrich Klinghardt, MD – physician practicing in the US and Europe and founder of the Klinghardt Academy of Neurobiology which offers in-depth training programs for health practitioners;

Henry Lai, PhD – Research Professor, Department of Bio-engineering, School of Medicine and College of Engineering, University of Washington;

Heather McKinney, PhD – researcher who specializes in strategic planning for effective change and health improvement within individuals, organizations and communities;

Rob Metzinger – Electronics Engineering Technologist, and instructor and consultant with the International Institute for Bau-Biologie® (IBE);

Alasdair Philips – the Director of Powerwatch (UK), qualified in Electrical and Electronic Engineering;

William Rea, MD – former surgeon who founded the Environmental Health Center in Dallas, Texas;

Hans Scheiner, MD – physician practicing complementary medicine in Germany, has treated thousands of electro-sensitive patients in his medical clinic;

Peter Sierck – industrial hygienist with IBE and IBN certification who has more than twenty years of experience performing professional EMF and RF measurements;

Stephen T. Sinatra, MD, FACC, CNS – board-certified cardiologist, Fellow in the American College of Cardiology and founder of Heart MD Institute, who lectures worldwide and writes a newsletter entitled *Heart, Health & Nutrition*;

Louis Slesin, PhD – founder and publisher of Microwave News, the newsletter which was established in 1980 and is now online;

Sarah Starkey, PhD – neuroscientist in the UK with a specific concern about wireless systems in the school environment;

Jacob Teitelbaum, MD – author of *From Fatigued to Fantastic* and a specialist in complementary medicine.

I have had the opportunity to communicate directly with these professionals and have benefited a great deal in rescuing my own family from everyday radiation exposure.

I have drawn upon the work of others through the available public information. Please keep in mind that this advice is based on the best information at this time, in a rapidly changing field.

In Step 1 these contributors, and many others, present the evidence which may have been hidden from us.

From a personal interview with Associate Professor Olle Johansson of the Karolinska Institute, and Professor of the Royal Institute of Technology, in Stockholm, Sweden:

> I urge responsible persons to think seriously about the full-scale experiment which is now in progress with us as the laboratory animals.
>
> I am particularly concerned about teens who are using mobile phones so much, sleeping with them on standby, and using them for alarm clocks.
>
> I am also concerned about parents using mobile and cordless phones, near babies and young children. There is evidence of harmful effects.
>
> You asked me about my hopes for the future?
>
> My hope is that I am 100% wrong.

The voluntary exposure of the brain to microwaves from hand-held mobile phones is the largest human biological experiment ever.

Professor L. Salford, MD, Lund University, Sweden

Step 1. Know the Evidence

- the levels of electro-magnetic radiation that we are exposed to today are billions of times higher than for our ancestors (yes, *billions*, and this is a conservative estimate)

- this technology was never pre-market tested for safety

- the current government standards we assume are protecting us are, evidently, just plain wrong

- **fortunately, there is good news** - this is not an insurmountable problem; there are safer solutions and hope on the horizon.

Facts do not cease to exist because they are ignored. Aldous Huxley

Microwaving this Blue Planet

As if satellites hovering over us in space and city-wide wireless systems weren't enough, there's now talk of wide ranging wireless networks for the entire USA and parts of Canada – a truly horrendous blitzkrieg?

You may be wondering, if scientists say there's evidence of harm, then why haven't our governments done something? And, how did wireless devices make it into the market in the first place? Good questions. Let's look at that.

Thermal and Non-Thermal – We Need To Know This

Way back in the techno dark ages – the early nineties – research on cell phones indicated the only health effect that could follow from exposure had to do with thermal damage – heating. Because cell phones did not heat tissue, the industry argued, they were safe and should be excluded from pre-market testing.

The safety standards were set by the International Commission on Non-Ionizing Radiation Protection, known as ICNIRP. You'll hear more about this agency as it seems they are at the center of quite a heated – thermal – debate.

Dr. Martin Blank, who is one of the world's leading researchers in this area, explains the thermal levels used in the ICNIRP guidelines are "a dead issue, scientifically." The trouble is, this is still the only thing that our government regulations are based on.

Check out the chart for a more detailed explanation. It may be too much detail for you right now, but it clarifies a complex issue and lets you see the two sides to this story. Also you need to know how the standards were set to see that there is not a slight difference in opinion on what levels are actually safe – they are quite literally thousands to millions of miles apart, as you will see in another chart up ahead.

And, as you read more of this section, you will know what the "thermal/non-thermal", "irrelevant safety standards" talk is all about. Having a basic understanding of this is very helpful.

Professor Johansson added:

> The only thing I should state more clearly (you have written about it, yes, but it may be a little bit 'hidden' for a non-specialist) is the difference between acute versus long-term effects. The current standards only look at acute thermal effects.

A 2-page Primer on the Regulations and Safe Levels Dispute

How can the regulations, the 'safe' levels, be so far off course, according to many?

To make sense of this issue it's crucial for us to grasp the basics. Please read this.

Points of View:

Regulators – who set electro-magnetic radiation (EMR) safe levels.

Concerned scientists/researchers – who are well versed in the biological evidence.

The Issue – Non-thermal vs. Thermal:

Non-thermal – emitted by most wireless and wired devices – in normal use will not heat tissue.

Thermal – the ability to heat tissue. Thermal levels of microwave/radio frequency radiation will cause harmful biological effects because they are high enough to heat tissue.

Note: thermal levels are the only measuring stick used in current regulations.

Step I. Know the Evidence

The Dispute:
Non-thermal levels – too low to heat our brains, eyes, etc. so must be safe?

Regulators' position: cell towers, mobile phones, wireless internet, power lines, power tools, household appliances etc. do not emit high enough levels to heat human tissue, so there is no cause for alarm. They do not recognize the evidence of biological effects on living tissue at much lower – non-thermal – levels of exposure.

Concerned scientists/researchers are calling for biologically-based standards. The current 'safe' levels are not based on biological effects and are not considered to be safe or accurate. Those standards, set in 1998, focused solely on damage caused by increases in temperature and did not consider other effects: damage to DNA, leakage of the blood-brain barrier and other biological measures of cell function and precursors to disease.

There is now clear evidence of damage that occurs at levels millions of times lower than existing standards. Just because your mobile phone can't cook your brain in three minutes, does not mean it is not causing harm. Brain cancer can take 10 – 20 years to develop. Damage to our DNA and other cellular dysfunction, headaches, dizziness, heat/pain on the side of the head, tinnitus, sleep problems, electro-sensitivity and neurological, immune and cardiovascular symptoms can occur *much sooner*.

A Bit Of History

By 1993, earlier studies were called into question and the American Congress, the FDA, and the cell phone industry struck a deal that the industry would not be regulated until new research was completed. The industry hired Dr. George Carlo. Dr. Carlo has training in medical science, pathology, epidemiology and law. He was given the job of overseeing a large research program.

He explains:

> Between 1993 and 1999, with more than 200 doctors and scientists from around the world participating, with the Harvard School of Public Health involved in peer review, and more than 56 studies conducted, we ran what still remains the largest program ever conducted in the world on the dangers of mobile telephony and wireless communications in general.

The team wasn't expecting to find anything significant. Evidently, there were multiple levels of peer review, studies duplicated in at least two labs, and the involvement of every major U.S. government agency involved in public health, safety and communications.

Dr. Carlo comments on the discovery:

> When we finished our work we found that indeed there were non-thermal effects that we had observed – making the government safety levels totally irrelevant.
>
> We identified the presence of genetic damage in human blood cells exposed to radiation from cell phones. We also had data that confirmed that there

were cellular dysfunctions, including leakage in the blood-brain barrier.

Most surprising was that we found near tripling in the risk of rare neuroepithelial tumors, rare brain tumors, in people who use cell phones when compared to people who did not use cell phones. It was a significant increase in risk, and those tumors were statistically significant in that they were correlated to the side of the head where the people reported using the phone.

Dr. Carlo and his team strongly recommended that, given the number of red flags raised by their research, the public should be informed of the results so that people could take precautions if they wanted to. Reportedly, the industry executives declined.

Remember the headlines "Go ahead and talk all you want—it's safe!"

When I try to tell people about this, most don't want to believe it – and that's not just the 'crackberry' addicts. People often point to a well-publicized, and controversial, Danish study released in December 2006. I have learned the importance of having a critical eye when reading such research papers. In the case of this Danish study, scientists not working for the industry found serious flaws. Evidently, the trouble with many studies that show no increased risk with cell phone use is that significant information may be missing. Here's another example to scrutinize.

In December 2009, a Danish study that reported, "No increase in brain tumors seen from cell phones" was widely repudiated by experts in this field. But how are we, the consumers, to know?

Independent researchers report that studies don't always ask what side of the head the person uses the phone on – a key variable in assessing risk. People are included as users even if they only had rare or occasional exposure. Also, only a very small proportion of the population had cell phones 10-15 years ago, and it can take 10 years or more for a tumor to develop. These studies wouldn't necessarily provide clear cause and effect.

How many decades, and thousands of studies, did it take before it was "proven" that smoking caused cancer?

If it can take so many years to see evidence of the damage what does this mean to a whole generation using cell phones day and night? Is it prudent to wait?

Those saying there's no problem, or who do not take precautionary action – in the face of growing signs of harm – may have a lot to answer for, one doctor confided in me.

Dr. David Servan-Schreiber, Professor of Psychiatry at the University of Pittsburgh, and Dr. Annie Sasco who worked with the World Health Organization agree that there are cell phone health concerns:

> What we have now are highly suggestive results from a number of different studies…that point to a non-negligible long-term risk for a brain tumor.

An Appeal From 20 International Experts by Dr. David Servan-Schreiber is on our website: www.radiationrescue.org.

Dr. Magda Havas, from the Environmental and Resource Studies Program, Trent University, confirms other risks:

> Headaches, numbness and tingling sensations in their fingers and face, and difficulty with cognitive functions either when using, or being in the presence of, cell phones and/or cordless phones. Also, people who live within 400 metres (437 yards) of the antennas are experiencing adverse health symptoms. Studies show an increase in cancers and symptoms of what is being called electro-hypersensitivity or EHS.

Why hasn't this been front page news for years? Another researcher, Raymond Paul Doyon, MA, has put together an extensive list of studies, and he wrote:

> The cell phone industry is in a catch-22 situation, if they warn consumers of the risks, and/or take visible steps to make cell phones safer, would they be admitting what they have been denying all along – that cell phones can be dangerous? Would this open them up to lawsuits?

Dr. Carrie Hyman is dedicated to educating lay and medical communities, as well as legislators. She offers this concern:

> Potential liability issues, rather than a clear concern about public safety, seem to be motivating the wireless industry's denial and obfuscation of scientific evidence of harmful effects of EMF (much of it provided by scientists who lost funding once evidence of harm showed up in their research.)

Dr. Henry Lai is one of a number of credible scientists who have faced attempts to discredit their findings.

The trouble started when Dr. Lai's work at the University of Washington showed that radiation at frequencies similar to those emitted by cell phones broke the DNA in rats' brain cells after just two hours of exposure (Lai & Singh 2004).

In "The Cellphone Game", an excellent and disturbing article published in the September 2008 edition of The Walrus magazine, Dr. Lai reports:

> Industry-funded studies are roughly twice as likely as government-funded ones to conclude that cell phones are harmless. (see sidebar next page)

Here is just one of Dr. Lai's several summaries of studies, with the researchers' name(s), and year of publication.

Studies on Wireless Communication-related Signals - Genetic Effects

Those done by independent researchers are shown below in plain text and **those funded by industry are in bold.**

Studies that show effect:

Aitken (05); Belyaev (05, 06); Czyz (04); d'Ambrosio (02); Diem (05); Gadhia (03); Gandhi (05a,b); **Goswami (99)**; Harvey & French (00); **Ivaschuk (97)**; Maes (96, 97); Markova (05); Mashevich (03); Nikolova (05); Pacini (02); **Phillips (98)**; Sarimov (04); Sykes (01); **Tice (02)**; Wang (05); Zotti-Martelli (05)

Studies that show no effect:

Antonopoulos (97); **Bisht (02)**; Chang (05); Chauhan (06a,b); Finnie (05); **Fritz (97)**; Gorlitz (05); **Gos (00)**; **Hook (04a)**; **Kuribayashi (05)**; **Li (01)**; Maes (01, 06); **Malyapa (97)**; McNamee (02a,b, 03); **Morrisey (99)**; **Qutob (06)**; **Sakuma (06)**; **Stronati (06)**; **Takahaski (02)**; Verschaeve (06); **Vijayalaxmi (01a,b, 03)**; **Whitehead (05,06)**; Zeni (03)

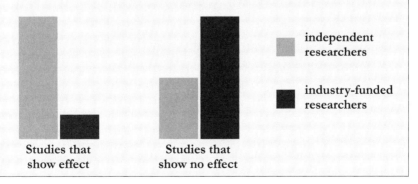

Step 1. Know the Evidence

The Interphone Study – Sorting Out The Science

In April 2009, Dr. Louis Slesin, editor of Microwave News, gave me his perspective on Interphone – an international study you will hear about. I asked him for clarification as there seems to be some disagreement among the researchers about the results.

My concern was that we, the consumers, might get the same kind of media spin we received with the Danish study we've just discussed. The "insiders" know how to sort out the science, but what about us?

(As you may know, an epidemiological study looks at data – the number of people who were exposed, the number who got sick etc.)

Dr. Louis Slesin explains some of the controversy:

> This epidemiological study began in the late 1990's with 13 countries participating. The problem is that although it is close to four years after the paper with the final results was first drafted, only partial findings have been published by some of the countries and some sub-groups. Finally at the end of May 2009 the Interphone paper on the possible link between mobile phones and brain tumors was submitted. There is enough evidence pointing to a tumor risk to prompt taking precautionary steps, according to respected Australian epidemiologist Professor Bruce Armstrong. Some of the other participating researchers have also come out clearly for precaution, while others say the results are inconclusive, so this will not be the final word. It is baffling how some

of the members of the team can be so sure that the phones are safe and give an all clear. Three different types of tumors have been implicated following ten or more years of cell phone use. No one should forget that the wireless industry plays a huge role in this area of research.

In October 2009, I asked Dr. Slesin to give us an update for this second edition, as he can offer a 'behind the headlines' story we may not always get from the media. He reports:

> The take-home lesson about cell phones continues to be that we should use caution, especially when it comes to children. Briefly, we have two different groups - a number of the Interphone study teams and Lennart Hardell's team - both saying that there is statistically significant evidence of two different kinds of tumors from cell phone use of ten years, or more, on the same side of the head as the phone was used. Of course, we are all waiting for the final analysis but there is already enough highly suggestive evidence to urge precautions.

As the focus of this book is seeking safer solutions, I welcomed Dr. Slesin's further comments, "You don't have to give up your cell phone. Use hands-free sets, the speakerphone - just keep it away from your head."

(You will see a list of recommendations in Step 3, including limiting use of all mobile devices, and using corded landline phones and wired Internet access to reduce exposure and risk.)

Let's hope someone is going to measure the potential health risks of hours a day – not minutes a month – of radiation exposure from the world's most popular mobile. As you may know, this "must-have" gadget now offers endless convenience and entertainment surfing the net, watching the news, downloading movies, and enjoying thousands of other applications. Remember, this is just during your waking hours; add to that the all-night exposure when using this mobile as an alarm clock.

And Lloyd Morgan, an electrical engineer well-versed in this issue, reminds us about cordless phones, "The Interphone study does not include the risk of brain tumors from cordless phone use. However, independent studies have consistently shown that cordless phone use has the same risk of brain tumors as found with cell phones."

Dr. Kelsey, we need your courage again

A tragic example, many years ago, of something that consumers assumed was safe, was a drug called thalidomide. Throughout many parts of the world, in the 1960's, 10,000 babies were born with shortened arms and legs, or with no limbs at all. Their mothers had taken this new prescription drug early in their pregnancies to help them sleep, and to keep them from feeling nauseated. Tragically, no one realized that thalidomide could cause terrible birth defects.

The United States was largely spared this disaster because Dr. Frances Kelsey, a Canadian physician was an alert reviewer at the Food and Drug Administration (FDA). Dr. Kelsey was not convinced that the manufacturer of thalidomide had proven that the drug was safe. She refused to clear the drug for sale in the U.S. Her caution paid off. Of the 10,000 babies injured worldwide, only seventeen were born in the USA. President John Kennedy honored Dr. Kelsey for her courage under fire.

Source: Office of Science Education, National Institutes of Health

Agencies Passing the Buck?

Unfortunately, the dispute about the regulations controlled by ICNIRP, the World Health Organization, and various federal government agencies, continues. This controversy is not very helpful when you're wondering about that powerful multi-purpose PDA you carry next to your body.

I heard one expert warn that these powerful new multi-task phones are getting close to thermal levels. Is some government agency monitoring this? Or monitoring the people getting symptoms?

Last year, I discovered that the European Union's main environmental watchdog, the EEA, and the BioInitiative Working Group – an international team of scientists, researchers and public health policy professionals – were both warning that government standards are woefully inadequate. Truth is, they're irrelevant, according to the BioInitiative Report.

(Check out www.healthandenvironment.org for some excellent information; if you want to read the BioInitiative Report go to www.bioinitiative.org.)

In January 2009, Dr. Slesin commented on these regulations:

> The Finnish Radiation and Nuclear Safety Authority has just recommended that parents limit their children's use of mobile phones and, on the same day, the French government announced a series of environmental health proposals, which include a ban on advertising that promotes the use of cell phones among those under 12...

> In contrast to all this activity, the lackadaisical approach of the International Commission on Non-Ionizing Radiation Protection (ICNIRP) is striking.
>
> Some 40 countries, many of which have only limited expertise in RF radiation health effects, look to the Commission for advice. **Yet, ICNIRP has been silent for ten long years. Their last guidelines on RF exposures were published in 1998.**
>
> How long could it take to write a simple statement urging caution?
>
> www.microwavenews.com

A European gathering was held in Brussels in February 2009 and 150 members from governmental agencies, advocacy groups, journalists, industry representatives and expert advisors to the EU science committees attended.

The Mobile Manufacturers Forum (a lobby organization) presented reasons why it believes its members are sufficiently protected from criticism and potential litigation – because products (i.e. mobile phones) comply for the most part with ICNIRP safety limits.

David Gee, one of the authors of the BioInitiative Report, told the audience that:

> mobile phone manufacturers will not be able to claim immunity if, in ten years or so, it is determined that use of mobile phones causes brain cancer or other illness.

> The mobile phone manufacturers are forewarned now, and simply saying 'but we were in compliance with ICNIRP safety limits' will not be sufficient reason for ignoring the growing evidence.

US Federal Regulations – Protecting Americans?

Professor Devra Davis, the noted epidemiologist comments:

> The agency that offers recommendations on cell-phone emissions in the U.S. – the Federal Communications Commission – doesn't employ a single health expert. The standards the FCC adopts are based on advice given by outside experts, many of whom work directly for the cell-phone industry. The Food and Drug Administration lacks the authority to set standards for cell phones and can only act if a phone is shown to release hazardous signals.

Yes, this is a lot of information to absorb and much of it may be new to you. I hope you are not reading this on your PDA.

We've seen how the regulations are not protecting us. As you read through this, please remember that help is on the way.

Scientists including Dr. Davis and Dr. Blank, and effective advocates including the publisher Dr. Louis Slesin, are gaining steam, and media coverage; people are waking up; legislators in the US, and other countries, will be forced to take action.

These are *our* families at risk.

Never underestimate the protective dedication of parents, and grandparents, when we realize our loved ones are in danger. We are not blinded by profits, convenience or shiny new electronics.

We will get international regulations based on the science to prevent harmful biological effects.

As you read the next section, with details on how this radiation may affect us, you will see why this is such an urgent call.

Do you read the fine print on your cell phone contract?

Who does? But you might check to see if you are agreeing not to participate in any class action lawsuits. Some have this clause.

Another interesting development is that after 2002, the insurance industry began excluding health risk claims from the product liability policies that they were selling to the mobile phone industry. In other words, it appears that insurance underwriters such as Lloyds of London may have refused to insure cell phone manufacturers against health-related claims.

Evidently, if the industry loses any of the lawsuits that have been brought against it around the world, bankruptcy could result. Remember how few class action suits it took to seriously damage the asbestos and the breast implant industries?

However, we are told that since as much as 30% of retirement fund shares are invested in the telecommunications industry, at least in the U.S., it could be disastrous to retirees, and by extension to the already bruised American economy, if the telecommunications industry went down in flames.

The friendly skies?

After the memorial service for her husband – her high school sweetheart – who died of a brain tumor, the widow is consumed with grief. He was a commercial pilot who had a cell phone for many years. One year she gave him one of those in-the-ear wireless headsets as she heard these reduce the radiation.

She discovers that the aircrew's occupational radiation exposure has caused concern; and that there is more exposure on the horizon. She read that a group of European physicians has issued an urgent medical alert about cell phones and wireless Internet onboard.

These physicians warn that this radiation can affect cognitive function and reduce reaction time – crucial factors in aviation safety, as you can imagine, and can put the crew at greater cancer risk. Surely, the strictly controlled aviation industry and other regulatory agencies are vigilant about the health of the crews and passengers.

Biological Effects - Evidence of Damage

(The Resources section at the end of the book also has an extensive list of the effects, with the scientific references. There *is* sound science.)

To the normal functioning of our cells

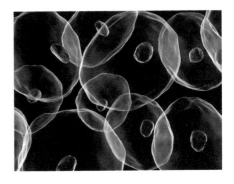

Scientists now have far more sensitive measures of how human cells respond to EMR than they did in the 1990s.

As the research of Dr. Blank demonstrates, these fields can affect our cells even when present at very low intensities – low enough that our senses can't detect them, let alone feel any heat. (Could this be the crux of the problem?)

Along with Dr. Reba Goodman, Dr. Blank has been studying these biological effects – near an electric clock, for example – on the stress response in cells:

> **The cells react as if this energy is harmful!** They make stress proteins only when they sense they're in a bad situation, such as when they are exposed to heat, toxins or the wrong pH.
>
> The research is showing that when exposed to low frequency they're responding in this way, even without thermal stressors. When the cell starts making stress proteins, it's trying to tell you something!

To Our Immune & Other Systems

Another scientist, Dr. Ted Litovitz, who was Emeritus Professor of Physics and Bioelectro-magnetics, Catholic University of America, Washington, DC, found that when a cell is subjected to too much stress, this fundamental protective response itself becomes stressed and stops making proteins. (Litovitz 2002)

In other words, the cell becomes less able to protect itself and this impairs our immune system strength and affects every system in the body.

To The Health of Our Cells and Their Ability to Communicate

The pioneer in the field of environmental health, Dr. Neil Cherry of New Zealand, described this mechanism in detail. Sadly, this highly respected scientist died in 2003 after suffering from Motor Neuron disease.

Professor Cherry wrote:

> Nature has developed many advanced and complex biological processes in, and between, cells for cell-to-cell communication, intra-cellular communication and regulation, and brain and CNS systems.
>
> Dr. Ross Adey's description of cells 'whispering to each other' in this cell-to-cell caring society, checking on the health of their neighbours and suggesting subtle changes to keep them healthy are part of biological homeostasis.

Our brains and cells primarily use calcium ions for these processes and Dr. Adey's laboratory, followed by Dr. Carl Blackman's monumental work, confirms that the natural and vital processes for cellular health are interfered with ... at the cell membrane level, by the oscillating signals from mobile phone base stations.

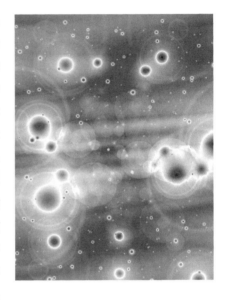

These waves of radiation confuse and damage the cell's signal system while the strong emissions of a cell phone next to your head almost drown out the signals. This alters the EEG and hormones damaging the cells and producing acute symptoms such as headache, concentration problems, memory loss, dizziness and nausea, and long-term problems including brain tumors, Alzheimer's Disease, depression and suicide.

The cardiologist Dr. Stephen Sinatra told me about this hypothesis:

Receptor sites on the cell membrane do not recognize the chaotic electrical disturbances from mobile phones and other wireless exposures, and interpret this as a foreign invader, sending the cell membranes into a protective 'lockdown' mode.

Evidently, this impedes the active transport channels that should allow nutrients into the cell and waste products to get out, and the cells' ability to communicate and work well with each other. Waste products can build up inside the cell causing a high concentration of molecules called free radicals.

To the DNA

Free radicals interfere with DNA repair and alter future DNA. Damaged DNA is passed along to the next generation of cells. This is worrying.

Because the cells can no longer 'whisper' with each other, they can no longer call in the immune system to remove dead cells and damaged DNA.

Another global alert: This disruption to the normal functioning of cells affects all life forms on this planet, not just us humans.

One of Dr. Hyman's presentations is titled *The Inconvenient Truth About Convenient Technology*. She cautions us:

> This radiation affects every living cell in ways we are only beginning to understand. Because of the genotoxic effect of radio frequencies in particular, we may be altering the genetic expression of all life on earth.

To The Cardiovascular System

Having worked for many years in the field of heart health education, I welcomed Dr. Sinatra's contribution to this updated edition and peppered him with questions about the possible connection between EMR and cardiac symptoms. Even if you have never had any worrying cardiac symptoms, you'll find what he says of interest:

> The heart is electrical in nature so it is vulnerable to all electro-magnetic fields. These include the earth's natural forces – as strange as this may sound, ask any cardiologist who has worked in the ER on a full moon night – as well as the microwaves from wireless technologies. The non-harmonious, jagged waveform of wireless radiation impairs the normal functioning of the heart, including the rate and rhythm. Some people can feel discomfort in the chest area when around mobile phones or wireless hot spots. People are usually unaware of the cause of these 'unexplained' cardiac symptoms, and unknowingly, their doctors may not offer the appropriate diagnosis, or treatment. Even if you are not aware of any effects, it does not mean that you are safe from harm.
>
> Cardiac effects may include: variability of heart rate and rhythm (tachycardia and arrhythmia), palpitations, chest discomfort/pain (angina), high blood pressure (hypertension - this is now better understood: oxidative stress/free radical damage from toxins, environmental pollution, emotional

stress etc. impairs the delicate lining of small blood vessels, causing constriction and an increase in blood pressure), blood clumping (this thickening of the blood also happens with the stress response, as you can imagine, thicker, more viscous blood is not a healthy condition).

Leakage in the Blood-Brain Barrier

Even though this is a complex mechanism, this research caught my attention – like the graphic on the back cover. I read that the "lockdown" effect of the cell membrane by EMR can affect cells anywhere in the body – including the blood-brain barrier, a vital filtering system in the blood vessels in the brain.

Experts agree that it's especially important that the cells of the blood-brain barrier keep toxins out and cerebrospinal fluids in, so that the environment of the brain is kept clean and the brain itself is cushioned against any contact with the skull. Makes sense.

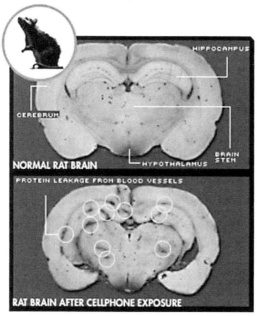

Research by Leif Salford, MD, the neurology professor at Lund University, shows that EMR is capable of causing leakage of the blood-brain barrier. (Persson and Salford 1996; Salford et al. 1992, 1993, 1994, 1997b, 2001)

Step 1. Know the Evidence

Dr. Salford's research team reported rat brain cross-sections showed first ever evidence of brain damage from cell phone radiation. While the controls (top) appeared healthy, the test subjects (below) – exposed to a 2-hour dose of cell phone radiation of varying intensities – were heavily spotted with proteins leaked from the surrounding blood vessels, and show signs of significant neuronal (brain) damage.

I once showed this graphic to a neurologist. He looked right at me and challenged: "Do you have any idea of the implications of this kind of damage?"

"I am beginning to understand," I replied. "This is why I am so committed to getting this information out to people – particularly to parents like me."

Maybe you're thinking that what applies to rat brains doesn't necessarily apply to humans? In 2008 Professor Salford pointed out that the blood-brain barrier is anatomically the same in both species and cautioned:

> With a long series of significant effects demonstrated in the animal models, it is my sincere belief that it is more probable than unlikely that non-thermal electro-magnetic fields from mobile phones do have effects upon the human brain.

So why are we flooding schools with wireless radiation?
I challenged one skeptical principal:

"If the safety of the school's playground equipment was being questioned by numerous highly respected scientists, and there were reports of children being harmed, and at risk, how long would it take you to remove it from the school grounds?"

Neurological Conditions

Reportedly, there are many symptoms connected with EMR exposure including difficulty sleeping, headaches, dizziness, eye pain, cardiovascular reactions, and many more.

Sadly, medical researchers also suggest an extensive range of brain and nervous system-related conditions.

These include: learning disabilities, ADD and ADHD, Asperger's Syndrome and the Autism Spectrum Disorders, MS, ALS, Alzheimer's and Parkinson's.

According to the neuroscientist Professor Johansson:

> If you look in the literature, you have a large number of various effects like chromosome damage, impact on the concentration capacity, decrease in short term memory and increases in the number of cancer incidences.

Many years ago, several studies reported a link between electrical occupations and neuro-degenerative disease including ALS (Suavity et al, 1998). I know several people afflicted with this devastating condition, and some with Parkinson's – otherwise healthy men in their 50's and 60's – all worked in the computer industry. Worth investigation?

There is also concern about the rising levels of Autism Spectrum Disorder (ASD). More and more families are facing this every day: a few decades ago autism struck 1 in 10,000 children; now it affects 1 in every 150, according to a study released in February 2007 by the CDC (Centers for Disease Control and Prevention). In October 2009, a new government study estimates that the prevalence is more likely about 1 in every 91 children.

There is some good news; as an environmental consultant who assists families with autism tells us later on, many of these children improve significantly when in a low EMR environment.

Too Scary?

A few years ago I submitted an article to a local parents' magazine. It was rejected as "too scary". Don't you think we should deal with this issue, instead of ignoring it? A head in the sand has never seemed safe to me.

As we've been hearing, credible research indicates damage from wireless communication can include brain tumors – our risk increases with time spent on cordless, or cell, phones; breast cancer – this tissue absorbs the electro-magnetic radiation increasing the risk; memory loss and concentration problems – the radiation negatively affects the ability of the brain to process and remember information, and dementia – the radiation can damage and kill brain cells, most of which do not re-grow.

We may not want to hear this, however, sadly, we need to know it.

As you read this, it may not seem obvious, but I am mindful of trying not to overwhelm you. Maybe it's time to put down this book and go for a brisk walk in the fresh air with your family?

My hope is that you will see that if environmental influences are part of these diseases, then we can prevent, and treat them, more effectively. The wellbeing of our families is precious. We have a whole generation at risk.

Childhood Cancers

Yes, this is scary, and what we are trying to prevent. I just saw a young child (not the one in this picture) in a magazine article about childhood leukemia. A little boy, in a grown-up looking suit, was holding a child-size cell phone.

In Australia, the UK and Italy, studies found that the closer people live to radiation from cell tower/broadcast antennas, the higher their incidence of leukemia (Michelozzi 2001, 2002, Hocking et al 1996, among others).

Other studies looked at people living near power lines. They discovered that with a small increase in the number of power lines, there was a doubling of childhood leukemia – by age 3 or 4 (Draper et al 2005).

For 30 years, experts have been trying to figure out exactly how electro-magnetic radiation could increase the risk of childhood leukemia, and in December 2008, medical researchers in Shanghai released a potentially ground-breaking explanation.

The researchers discovered children who carry a defective version of a gene, that normally helps repair damaged DNA, are up to 4 times more likely than neighboring children, with a fully-functioning gene, to develop leukemia, if they also live near power lines or transformers.

They report that children who have this faulty gene cannot repair DNA that's damaged by exposure, even to low frequency fields, which would make them more susceptible to cancer.

There's also a possibility that a mother's exposure in her work environment during pregnancy, or just before she conceives, will also put her child at greater risk of developing certain types of brain cancer.

In February 2009, researchers at Canada's McGill University raised this concern after following the pregnancies of a group of women working in a room full of electric sewing machines.

With these serious implications wouldn't you think that government health agencies and research institutions would be leading the charge to prevent these side effects?

So this might become the new health warning during pregnancy: refrain from drugs, alcohol, tobacco, prolonged proximity to electric devices, and all mobile technology – cordless and cell phones and everything wireless.

As with tobacco, and other public health issues, there's usually a long time lag between people getting sick and governments taking action, as we mentioned. We need to remember this.

Not only are the agencies in charge not stepping in to fund needed research and provide necessary precautions, most are still defending the existing 'safe' levels which respected scientists warn are not protecting us.

Meanwhile, childhood brain cancers are on the rise.

In April 2009, the British neurosurgeon Kevin O'Neill, MD reported:

> Brain tumours are on the increase, reportedly in the region of 2% per year. But in my unit we have seen **the number of cases nearly double in the last year.**

A few years ago, I heard some scientists and epidemiologists voice their greatest fear: the potential of a global pandemic of pediatric brain cancer. I shuddered and became even more determined to alert consumers. How much does the convenience really mean to us, when we consider such risks?

We feel harrowed seeing the bright face, and shiny bald head, of one child with cancer. How would we cope with many? How do families live through losing a child?

There are dedicated researchers and message carriers, around the world, fighting hard to prevent this. We need your help.

Dr. Davis is particularly concerned about our children:

> When it comes to proof of human harm, we've kind of been led down a garden path that says the only proof that really counts is enough sick or dead people. I think that approach has got to change, and particularly with respect to something like brain cancer.
>
> Brain cancer can take 10, 20 or 30 years to develop. It may affect people 30, 40 years after they've been exposed.
>
> Our children are using cell phones in growing numbers. They're marketing cell phones now to five-year-olds...
>
> Should we wait until we have proof of sick, or dead, kids before acting? I don't think so, and neither does the European parliament.

April 2009: Dr. Louis Slesin is another strong voice urging more research:

> Here's the killer: there is not a single study on cell phone radiation going on in the United States today. Nothing. There's a big study in the wings, but it hasn't started yet. But if that gets under way, we'll have one study. That's it.
>
> Now, with four billion people worldwide using cell phones, could we spend a few dollars finding out whether our children are at risk or not?

A partial summary of EMR biological effects

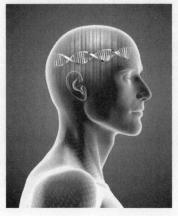

- DNA damage and disruption of DNA repair (Lai, Singh, Philips)

- Suppression of the immune system (Johansson, Nakamura, Litovitz, Veyret, Hocking, Draper)

- Disruption to normal functioning of neurological, cardiovascular and endocrine system (Lai, Salford, Becker, Cherry, Hurtado, Johansson, Karasek, Schilowsky)

- Leakage of the blood-brain barrier (Salford, Persson)

- Impaired cognitive functions (Mosgoeller, Scheiner, Lai, Becker, Cherry)

- Increased agitation, sleep disruptions and food, chemical and electro- sensitivities (Becker, Holt, Rea)

- Stress protein synthesis (body's reaction when stressed at the cellular level) (Blank, Goodman, Mosgoeller)

- Inter- and intra-cellular functions (cell membrane permeability and the cells ability to communicate with each other) (Cherry, Adey, Blackman)

More Evidence of Harm

Building Biology consultant and Industrial Hygienist, Peter Sierck, with his colleague Thomas Haumann PhD, offer this summary of non-thermal biological effects from wireless technologies. (Nonstop Pulsed 2.4 GHz Radiation Inside US Homes) You can show this to your friend (or government agency, or industry spokesperson) who protests, 'there is no real evidence of harm'.

"Much experimental evidence of non-thermal influences of microwave radiation on living systems has been published in the scientific literature during the last 30 years – relating both to in vitro and in vivo studies.

- Changes in the electrical activity in the human brain,
- Increase in DNA single and double strand breaks from HF exposure to 2.45 GHz,
- Increased lymphoma rates (2 fold) in transgenic mice exposed twice a day to 30 minutes of cell phone (GSM) signals over 18 months,
- Increased permeability of the blood-brain barrier in rats,
- Observation of an increase in resting blood pressure during exposure,
- Increased permeability of the erythrocyte membrane,
- Effects on brain electrochemistry (calcium efflux),
- Increase of chromosome aberrations and micronuclei in human blood lymphocytes,
- Synergistic effects with cancer promoting and certain psychoactive drugs,
- Depression of chicken immune systems,
- Increase in chicken embryo mortality,
- Effects on brain dopamine/opiate electrochemistry,
- Increases in DNA single and double strand breaks in rat brain,
- Stressful effects in healthy and tumor bearing mice,
- Neurogenetic effects and micronuclei formation in peritoneal macrophage."

The jury is still out? Or not?

There are many experts urging caution, and others, including an industry spokesperson, assuring us that there is "no problem as standards have been set with large safety margins".

How are we, the consumers, supposed to know who is independent, and who has close ties to the wireless industry, or has a need to stay on the path of least resistance – the status quo?

Have you ever heard Dr. Vini Khurana speak?

Dr. Khurana, MBBS, BSc (Med), PhD, FRACS is Staff Specialist Neurosurgeon at the Canberra Hospital, and Associate Professor of Neurosurgery at Australian National University Medical School.

On May 27, 2008, CNN's Larry King asked Dr. Khurana to back up his contention that the danger of cell phones could have far broader health ramifications than asbestos and smoking.

Dr. Khurana's reply:

> I base my statement on the fact, Larry, that at this point in time, there's just over three billion users of cell phones worldwide.
>
> So that's half of our world population, or almost half. In Australia, there are 22 million cell phones and 21 million people.
>
> And the concern is not just brain tumors, but other health effects associated or reported to be associated with cell phones, including behavioral disturbances,

salivary gland tumors, male infertility and microwave sickness syndrome.

And with so many users and so many young users, we should be concerned.

Does Your Family Fun Center Around Electronics?

I realize it may be hard to take in how we have been deluded for so long. Here's the reaction of one of my astute colleagues:

> Honestly, I felt myself wanting to turn away as you were telling me about your new book. I began to see how many of these gadgets are in my own wireless-filled home. My husband is almost proud of his 'crackberry' habit. Now I am wondering if this may be why he is getting headaches and dizzy spells. The doctors couldn't find anything wrong. Said it was stress. Seems it is another kind of stress.

Cell Phones For Their Safety, Video Games For Activity?

Many parents feel safer knowing their children/teens carry cell phones: you know where they are, and in case of an emergency they can call for help.

The kids text their friends, surf the Internet, and play an array of hyper-speed video games.

How many minutes, or hours, a day do your children spend gaming? Do you know?

The games that get them up off the couch and firing away at the screen are great exercise, some believe.

Parents may rationalise it's a real workout and keeps them off the streets. Unbeknownst to most parents, the kids are getting a lot more from this rapid-fire obsession, and many are becoming addicted.

Young people often sleep with cell phones under the pillow to catch that secret call, or text message.

Maybe this is why many have trouble getting to sleep and getting up for school? Who wants to sleep when you can twitter?

Dr. Hans Scheiner, a physician with a complementary medicine clinic in Germany told me:

> Children and young people can become <u>physiologically</u> addicted to EMR. So excessive cell phone use and wireless computer games are not just an addictive habit, or a behavioral issue.
>
> **EMR opens the blood-brain barrier in the same way as alcohol and drugs.** This can give a feeling of euphoria, and when you try to get them away from the game they'll experience withdrawal symptoms, as you may know if you've tried to reduce your child's computer/gaming time and cell phone use.

Children and Teens on Cell Phones and Cordless Phones

Consider this study from a leading researcher: in September of 2008, the Swedish oncologist, Professor Lennart Hardell, presented new evidence – an analysis of data from one of the biggest cancer risk studies – which shows that:

> children and teenagers are five times more likely to get a certain kind of brain cancer – glioma – if they begin using cell phones before the age of 20. The extra risk from using cordless phones was found to be almost as high.

When this story was published in the Canadian press, a spokesman for the Canadian Wireless Telecommunications Association told the public:

> Government agencies responsible for compiling and analyzing research – including Health Canada and the World Health Organization – continue to say that the evidence that is out there, that has been reviewed for years and years and years, says there is no demonstrated risk for human health.

Health Canada's response: "currently we see no scientific reason to consider the use of cellphones as unsafe. There is no convincing evidence of increased risk of disease." (Excerpted from an article by Sarah Schmidt, CanWest News Service)

Hardell's research shows risks also increased with the number of hours of use, the intensity of the phone's power, and the younger the child was when starting to use cell phones. As you know, the children of today will be exposed for decades longer than those of us who got our first mobile in our thirties or forties.

Studies from two other European labs, published in *Physics in Medicine and Biology* (June 7, 2008) show that young children's brains absorb twice as much radiation from a cell phone as the brains of adults.

Dealing with my children, and others who seem "hard-headed", it's difficult to believe but it turns out that children also have thinner skulls and smaller heads than adults, so radiation can penetrate more deeply into the brain matter – as you see on the book cover that shows an estimation of the absorption of electro-magnetic radiation from a cell phone based on age.

Dr. Davis explains:

> Few parents know that radio-frequency signals reach much more deeply into children's thinner and smaller brains than ours – a fact established through the pioneering work of Professor Om P. Gandhi, the leader of the University of Utah's electrical engineering department. The award-winning Gandhi worries that all the standards used for phones apply to the "big guy" brain.

> In 2004, standards became looser, because industry modelers decided to use a new approach – basically doubling the amount of radio frequency that could reach the brain of an adult and quadrupling that reaching a child's.
>
> The brain of a child doubles in the first two years of life and keeps on developing until their early 20s.
>
> Professor Gandhi no longer works with the cell-phone industry and none of his grandchildren uses a cell phone.

Dr. Carlo also points out:

> From a physiological point of view, children are more susceptible to damage because their cells are differentiating more.
>
> The consequences of disrupting cell differentiation are potentially devastating, making it a medical necessity to examine wireless networks in hospitals schools, homes, apartments, and communities.

In the spring of 2009, scientists in Europe and the USA are endorsing the BioInitiative Report and calling for world-wide health protection agencies and governments to urgently comply.

The following precautionary actions are proposed:

- **children under 16 should only use cordless and mobile phones in emergencies;**

- no Wi-Fi, WiMAX, or other wireless networking, be placed in homes, schools, or public areas; and

- conduct independent, widely publicized, frequent audits to ensure base stations do not exceed newly warranted, biologically based guidelines.

(Johansson, Olle, 2009. The London Resolution. *Pathophysiology*. In Press.)

How Can We Limit Our Teens' Risky Behaviors?

This generation is hard-wired to wireless gadgets. And most of their age group has never been prone to accepting adult advice.

When I was buying a pair of shoes the other day, the young sales person was working his cell phone every few minutes. I felt compelled to ask him if he had heard about health hazards.

The look he gave me wasn't quite the adolescent eye-roll I get from my own children, but it was certainly an expression of disbelief. "Not really interested", he shrugged, "I use this cell phone for everything."

I have consulted with Dr. Allison Rees, an expert in parent – teen communication for her advice on how we can approach this challenge; this is coming up with our Action Plan in Step 4.

Professor Olle Johansson offers some sobering thoughts:

> Children/teens may be more vulnerable due to the permeability of the skull and maybe the risk will turn out to be equally large for all ages.
>
> But, much more importantly, the effect on children is much more serious for mankind since it hits us at our 'roots' rather than at the 'end-branches', the elderly.
>
> As our knowledge stands right now, interaction with fertility, the immune system capacity, shattering of the DNA, or impaired learning and intelligence, will be much more dangerous than the extra cases of cancer.
>
> Of course, for anyone getting cancer, or being a relative to a cancer victim, they will suffer tremendously, no doubt about that. But, just considering mankind, as a species, we can lose a few, but not everyone.

I have just returned from the standing-room only 'Celebration of Life' of a truly marvelous man. Michael inspired everyone who met him with his humour, modesty and grace – before and after his diagnosis.

Even with a scar on the side of his bald head; even when he could no longer walk, he greeted each person, and each situation, with an ear-to-ear smile, radiating kindness.

He accepted each day with joyful zest, and always put others first, even toward the end of his journey when he was utterly dependent on them.

His wife with such courage, and his two young children with such promise, brought tears to our eyes and somehow hope to our hearts, as they remembered the love and laughter that he brought to them. Michael was 42.

Too much to contemplate?

Please remember this book is my compilation of more than three years of researching this. I have talked personally with many of these experts, made a lot of changes in my life, and experienced a lot of benefits. I have had time to try and come to terms with this wake-up call, and still find it hard to swallow and digest this information, so I empathize if you are choking on it.

Once more, Dr. Becker was ahead of his time:

> Since our civilization is irreversibly dependent on electronics, abolition of EMR is out of the question.
>
> However, we must halt the introduction of new sources, while we investigate the biohazards of those we already have, with a completeness and honesty that have so far been in short supply. New sources must be allowed only <u>after</u> their risks have been evaluated.

If Dr. Becker's advice had been followed we would not be facing the great challenge that lies before us today.

So who is taking action?

As you know, Europe and the UK are more densely populated than most of North America, and with that comes a correspondingly thick cloud of electro-smog. In some cities the ambient levels of radiation are so high that virtually everyone is being exposed, even those choosing not to use cell phones.

There are reports that in many countries upwards of 5 – 10% of the population is already experiencing related symptoms, so that may be why the wireless safety issue is getting more attention from legislators.

In September 2007, after an extensive study, the EEA – Europe's top environmental agency – called for immediate steps to reduce exposure from cell towers and other wireless radiation, warning that delay may result in a health crisis more serious than AIDS, and those caused by smoking, pesticides, asbestos and leaded gasoline.

In its report this agency advises us about standards around the world:

> Safety limits set for radiation are thousands of times too lenient, and an official British report concluded that it could not rule out the development of cancers.
>
> The EEA's initiative will increase pressure on governments and public health bodies to take precautionary action over the electro-magnetic radiation from rapidly expanding new technologies.
>
> The German government is already advising its citizens to use wired Internet connections instead of Wi-Fi, and corded landlines instead of mobile phones.

Safe Levels? A Difference of Opinion

Why do we need to know this? I want you to see the massive disparity between the levels considered biologically safe, and the ICNIRP safety standards that govern the marketplace.

Electrical Sources:

AC MAGNETIC FIELDS (Low Frequency, ELF/VLF) – from plugged in devices, household appliances, power lines, ground current issues. Tested with a Gauss meter (mG = milligauss)

Building Biology Guidelines (SBM 2008) for your sleeping areas

in milligauss	No concern	Slight concern	Severe concern	Extreme concern
mG	**0.2**	**0.2-1**	**1 - 5**	**5**

Daytime measures (taking into consideration the duration, proximity etc.) should be **below 1 mG**

Dr. Martin Blank: thresholds of biological effects 2 - 5 mG, so recommend less than that.

North America – ICNIRP regulations allow: 833 mG for the general public, and 4,166 mG for occupational exposure.

Europe – 1,000 mG for the general public and 5,000 mG for occupational exposure: Source: ICNIRP, 1998.

Wireless Communication:

RADIO FREQUENCY RADIATION

Cell towers, Radio, Television, DECT cordless phones, wLAN.

Test these fields and devices with an RF meter. Make sure your meter's frequency range is high enough – up to 5.8 GHz.

Some don't have a high enough range to test newer cordless phones, so you may get a false low reading.

Building Biology Guidelines (SBM 2008) for your sleeping areas

microwatts per square metre	No concern	Slight concern	Severe concern	Extreme concern
	< 0.1	0.1-10	10-1000	> 1000

Daytime measures should be: **below 10**

Dr. Martin Blank concurs RF threshold of biological effects: **0.1**

NOTE: ICNIRP levels allow 2 to 10 <u>million</u> microwatts per sq. metre

Step 1. Know the Evidence

In a few pages you will see the RF Guidelines table that shows international levels of this radio frequency radiation (also known as microwave radiation) from broadcast antennas, cell towers (masts), radar, satellites – wireless communications that now encircle all parts of our planet.

Remember what you just learned about thermal and non-thermal? Some of these international guidelines are based on the view that as long as the radiation is not powerful enough to cook tissue, then we're just fine. (Or, so their thinking goes.)

I asked the Building Biology environmental consultant, Katharina Gustavs, "just how high are those so-called 'safe' thermal levels?"

Ready for an astronomical number to stretch your comprehension? She answered, "At exposure levels of around 1,000,000,000 (microwatts per sq. metre in the frequency range of 400 MHz to 3 GHz), the eye lens starts to turn opaque; this is referred to as the thermal level."

Yes, that is one billion! So you see why some occupational exposures, including the military in many countries, allow 100,000,000 – why, it's only 1/10th of the allowable thermal level.

If you don't recognize the evidence of biological effects that have been shown to occur at a minuscule 0.1, it can seem that there's 'no problem'. This is the core of the issue, as I see it.

To put this into perspective, I also asked Peter Schlegel, the

Swiss civil engineer who created this table, "what was the natural background of RF radiation for our ancestors?" I had to pause and count the zeros when he replied, "0.000001 is the estimate."

How's that for an indication of how far we've come in a century of technological development? Schlegel goes on to explain, "this is the cosmic background radiation arriving on earth (source H.-P. Neitzke, ECOLOG Institute, Germany) and it is also a level that living cells respond to."

Knowing this is useful, he tells us, for raising the awareness of the harm we have caused to our environment. However, this low natural level does not exist anymore because it is overpowered by the emissions of our human-made technology.

From 0.000001 to 100,000,000 – anyone have a calculator? One presumes the various species living on this planet had millions of years with this natural background. You can see why so many experts are alarmed at the dramatically increased levels of radiation exposure we're forced to endure now.

The gap between what many scientists consider safe (0.1) and what some organizations allow (100,000,000) is the same difference in magnitude between the length of a few football fields and the distance from the earth to the sun. (See table, next page)

RF Exposure from Cell Towers (Masts)

Worldwide Limits and Guidelines (in microwatts per sq. metre - $\mu W/m^2$)

100,000,000	US Army, Navy, and Air Force; Bell Telephone; General Electric Company (1957/58) "Microwave Conference", USA 1955 [Brodeur, 1980]
10,000,000 to 4,500,000	USA, Canada, UK, Germany, Sweden, Finland, Japan, Austria, and many others. Switzerland total sum of RF at any place. Limit values depending on carrier frequency. ICES; ICNIRP: derived from biological effects of short-term, high-level exposures causing a rise in temperature.
2,000,000	Australia and NZ for GSM 900
1,200,000	Belgium without Wallonia Law (2001)
1,161,000	Italy, sum total of RF at any place
1,000,000	Former GDR, exposure ≤ 2 hours - OHS Regulation:TGL 22314 (1969)
100,000	Former GDR, exposure ≤ 20 hours - OHS Regulation:TGL 22314 (1969)
100,000	Former Soviet Union, mid-20th century
100,000	Italy, exposure > 4 hours
100,000 to 42,500	Switzerland, indoor exposure level from one transmitter site1. Limit values depending on carrier frequency. [1]A transmitter site may consist of one or more mobile phone base stations, its perimeter (ca. 40-100 m) depends on the radiated power of the particular adjacent base stations. Ordinance (2000).
24,000	Belgium: Brussels and Wallonia - Law (2007 and 2009)
1,000	Liechtenstein: indoor exposure level from one transmitter site; implementation by end of 2012 - Law and Ordinance (2009)
1,000	France, testing period decided by 60 cities - (Sep 2009)
1,000	Salzburg Resolution (1998), sum total of GSM
100	European Parliament, STOA Report - (2001)
100	BUND, prevention of hazards BUND – German Alliance for Environmental and Nature Protection (2008)
1	BUND, precautionary principle
1	Salzburg Precautionary Value, indoor exposure level for sum total of GSM - (2002)
< 0.1	**Building Biology Evaluation Guidelines** (SBM 2008) specifically designed for sleeping areas and empirically established on the basis of many thousands of individual cases, including EHS sufferers. Generally adopted by all Building Biology environmental consultants who conduct professional RF surveys in homes, at workplaces, for EHS sufferers.

0.000001 **Natural background** – mostly this low natural level does not exist anymore.

Prepared by Peter Schlegel, www.buergerwelle-schweiz.org based on a list published in "Warum Grenzwerte schädigen, nicht schützen [Why exposure limits are not protective, but harmful]" by the Competence Initiative www.broschuerenreihe.net/international/index.html (September 28, 2009). Edited by Katharina Gustavs. Printed here with permission.

An excerpt from Radiation Rescue (2010) by Kerry Crofton, PhD www.radiationrescue.org

How is this massive safety margin gap affecting our families? Do we even know? I have a young cousin in a navy radar unit; it is chilling to realize the extreme levels he is being exposed to onboard ship.

Now that you know about this measurement of RF radiation let's take a look at some of the levels we may be exposed to in our daily lives.

What is the level in your community?

Katharina Gustavs, who does these kinds of assessments, reports:

> For the sum total of the background RF radiation level in urban areas, you can find in the literature levels ranging from one hundred (microwatts per sq. metre) all the way up to a few thousand or even into the ten thousands, depending on how close the next base station is located.

This constant environmental exposure – electro-pollution – is why many dedicated advocacy groups are fighting the installations of cell towers/masts in their communities. You can find a group in your area, or start your own. (Check out The Statement of Accountability that is listed in the Index.)

Reducing RF Exposure Begins At Home

While it is essential that we have some comprehension of the exposures in our external environment, and that we make our voices heard, we should also be mindful of our exposures inside.

Another Building Biology consultant, Peter Sierck, reaffirms what many of our other technical experts have told me:

> We see many people fighting the installation of cell towers (masts) in their communities and then going back home where they have wireless routers at their workstation or desk, cordless phones which constantly radiate without being in use, microwave ovens in their kitchens, and often mobile phones at their ears.

Common Exposure Levels (all in microwatts per sq. metre)

Let's take a look at some measures taken by Gustavs, and some of our other assessment experts, in people's homes and workplaces. (There's lots more information about these coming up in Step 3.)

Wireless Internet access

An average-powered home wireless router
    Within one foot (30 cm) – 30,000
    within 10 ft (3 metres) – 3,000
Note: If you have a wireless router, how far is it from where you sit? Sleep?

A cordless phone (5.8 or DECT 6.0)

    Within one foot (30 cm) – 200,000
    within 10 ft (3 metres) – 2,000

Note: How far away is the cordless phone base station, or separate handsets, from where you sit? Sleep?

A microwave oven in use (new, not defective)

Within one foot (30 cm) – 10,000-1,000,000
within 10 ft (3 metres) – 1,500

Note: Remember this next time you, or your children, use a microwave oven.

In an average home size 2-3 bedroom home

– that has a wireless router and at least one cordless phone:

(KG) "It would have to be determined where the router and cordless phone are located. If the cordless phone is in the bedroom, the person in bed may be exposed to 20,000 all night long, and in the kitchen maybe only 200. Again, the router at the desk may expose the computer user to 20,000, while in the living room next door you may have 1,000, and in the bedroom further away – 100."

And at your school? Let's remember the children

A concerned mom who read the first edition of Radiation Rescue wrote to me asking what she could do about the cell tower antenna on top of the hill near her young children's school as the principal reassured her that, "The levels in the schoolyard were well within the government safety standards." Yes, we have heard that many times. I sent her this new table and suggested she show it to the school, and to other parents at the next meeting. Who better to fight for the wellbeing of our children than us, the parents?

Dr. David Carpenter, Director, Institute for Health and the Environment at the University of Albany and co-author of The BioInitiative Report warns:

> **This report stands as a wake-up call that long-term exposure to some kinds of EMF may cause serious health effects.** Good public health planning is needed now to prevent cancers and neurological diseases linked to exposure to power lines and other sources of EMF. We need to educate people and our decision-makers that 'business as usual' is unacceptable.

It is still an uphill battle, as you can imagine. In October 2008, Dr. Carpenter testified at a US Congressional hearing on the safety of cell phones, along with Dr. Ronald B. Herberman, an internationally recognized tumor immunologist.

They stated that according to the government's own statistics, the incidence of brain cancer has been increasing over the last ten years, particularly among 20-29 year-olds.

If it takes more than ten years for brain tumors to grow, and these are, in fact being caused by cell phones, cancer rates might not peak for at least another five years, in Dr. Herberman's opinion.

This cancer specialist advises:

> Recently I have become aware of the growing body of literature linking long-term cell phone use to possible adverse health effects including cancer.
>
> Although the evidence is still controversial, I am convinced that there are sufficient data to warrant issuing an advisory to share some precautionary advice on cell phone use.
>
> An expert panel of pathologists, oncologists and public health specialists recently declared that electromagnetic fields emitted by cell phones should be considered a potential human health risk.

Julius Knapp, from the Federal Communications Commission (FCC), also testified at the Congressional hearings, saying he doesn't have an opinion about whether federal safety standards need to be changed, because **"nobody at the agency is competent to evaluate the biology."**

As you may know, the FCC's job is to implement regulations.

Although it was invited to testify, the industry's trade group declined, and issued a statement assuring us that there's no scientific basis for any concern about cell phones. Getting skeptical of such assurances?

There are those, including government health regulators, who acknowledge that it is true that studies show EMF exposure causes biological effects, but still insist there's no proof that these effects lead to illness or disease.

Not so, according to Dr. Magda Havas, who says there are 'literally thousands of studies showing adverse health effects'.

At the time of this publication many researchers are voicing concerns about some electro-magnetic technologies – green and otherwise – and advising caution with SmartGrids and expanding wireless Internet networks. Note these broadcast communication data through open space radiating everything in its sphere. There are health, and security, concerns.

Many are proposing fiber optics for the wellbeing of all life forms and for greater security. Since this is a closed system, the data being sent along fiber optic cables are secure and there are no hazardous electro-magnetic radiation emissions.

As we mentioned earlier, the electro-pollution affects all life forms, not just humans.

Where have all the bees gone?

The beekeeper stares with disbelief and mounting concern: the bees, it seems, have disappeared again.

As he stands in his field inspecting the empty hives, he looks up and sees the cell tower antennas glaring down at him from the peak of the hill nearby. "Such unsightly things", he grumbles.

Half a world away, in an English village, an elderly church custodian is blissfully unaware of the cell antenna in his neighborhood; it is embedded in the shiny new metal cross, high upon the church tower.

He has been watching for the return of the barn swallows, as he has done for years. Where are the birds, he wonders? Why are they not making it back here?

Evidently, the adverse effects of disrupting the intra-cellular whispering that Dr. Adey described are not limited to the human species. Some researchers say EMR is also disrupting the natural navigation abilities of whales, birds and bees.

Dr. Carlo offers this explanation:

> The disappearance of bees – Colony Collapse Disorder – has occurred concurrently on four continents within a very short time frame.
>
> This hardly ever happened before and its implications are dire: how will our crops become pollinated? What will become of our food supply?
>
> If the reason was biological or chemical, as some say, there would be a pattern of epidemic spread – we would be able to trace the source, similar to the spread of SARS a few years ago. That is not the case.
>
> The colony collapse has hit each continent at roughly the same time. That would mean the cause hit the continents at the same time, as well. Mobile phones meet that criterion.
>
> The information-carrying radio waves from these, and other, wireless devices, disrupt inter- and intra-cellular communication in most species.
>
> Therefore, it is biologically plausible that the bees, exposed to this radiation, cannot find their way back to the hive, and starve.

Agriculture At Risk

Perhaps, we should consider this quote from Albert Einstein, "No more bees, no more pollination ... no more humans!"

It's Not Just The Bees

In 1971 William Kenton of Cornell University studied the effect of electromagnetic fields on birds' ability to migrate:

> ...with 1-gauss (electrically-charged) magnets attached to their heads the birds couldn't find their way home. Each bird without magnets faultlessly navigated the 150 miles ...and fluttered in like a helicopter to a perfect landing.

Arthur Firstenberg wrote in *The Ecologist* in June 2004:

> Canadian researchers working with parakeets, chickens, pigeons and seagulls reported that these birds avoided microwave fields (EMR) if they could, and collapsed within seconds if they couldn't. Many years later, Alfonso Martínez carefully documented the decline and disappearance of white storks, house sparrows, and free-tailed bats from the vicinity of cellular phone base stations.

More than forty-five years ago, Rachel Carson wrote:

> Over increasingly large areas of the US, spring now comes unheralded by the return of the birds, and the early mornings are strangely silent where they were once filled with the beauty of bird song.

More than 100 years ago, Anton Chekhov warned:

> Human beings have been endowed with reason and creative power. But they have been more destructive than creative. Forests are disappearing, rivers are drying up, and wildlife is becoming extinct.

Before a radar station in Latvia was shut down at the end of the Cold War, a co-ordinated effort was made to determine whether the station had any environmental impact.

Even at extremely low levels of exposure, researchers found these effects: smaller growth rings in trees, premature ageing in pine needles, (a factor in raging wildfires in heavily populated, high-EMR areas?) and they found chromosome damage in cows.

And Wolfgang Volkrodt linked forest die-back to microwave radiation rather than acid rain.

Also in Germany, Löscher and Käs documented illness in dairy cows caused by cell towers. You don't have to be a dairy farmer to realize the serious implications of this.

Wildlife and Domestic Animals At Risk

In 2009, wildlife biologist Alfonso Balmori, PhD of Valladolid, Spain reported that:

> **Electromagnetic radiation is a form of environmental pollution which may hurt wildlife.** Phone masts (cell towers) located in their living areas are irradiating continuously some species that could suffer long-term effects, like reduction of their natural defenses, deterioration of their health, problems in reproduction and reduction of their useful territory through habitat deterioration.
>
> Therefore microwave and radio frequency pollution constitutes a potential cause for the decline of animal populations and deterioration of health of plants living near phone masts.

Nature At Risk

Dr. Ulrich Warnke of the University of Saarland, in Germany cautioned a recent conference organized by Radiation Research in the UK that electro-smog was "disrupting nature on a massive scale". Dr. Warnke went on to say:

> **Today, unprecedented exposure levels from numerous wireless technologies interfere with the natural information system and functioning of humans, animals, and plants.** The consequences of this development, which have already been predicted by critics for many decades, cannot be ignored anymore.
>
> Bees and other insects vanish; birds avoid certain places, and become disorientated at others.

> Humans suffer from functional impairments and diseases.
>
> And insofar as the latter are hereditary, they will be passed on to next generations as pre-existing defects.

There's reason for hope

Some legislators and courts are taking action

There are encouraging signs that the tide is turning: some government agencies are listening to the independent scientists urging them to bring the regulations in line with the science.

Many parent and consumer groups are demanding product safety and accountability – people, and wildlife, over profit.

Proactive people all over the world are deciding not to wait for absolute proof before they take precautions, especially when it comes to the health of young, developing brains.

And while some governments are lagging behind, others are opting for the precautionary principle.

In the US

Representative Dennis Kucinich (Dem), who launched Congressional hearings into cell phones and health in 2008, says he will not let this issue drop.

In September 2009 I attended an international conference on cell phones and health, organized by Dr. Devra Davis in Washington, DC. One afternoon, we attended a US Senate Hearing on this subject, requested by Senator Arlen Specter of Pennsylvania and chaired by Sen. Tom Harkin of Iowa. Sen. Mark Prior of Arkansas also attended.

A scientist (speaking for the wireless industry?) caused a stir in the gallery with the claim, "The current scientific evidence does not demonstrate that wireless causes cancer, or other adverse health effects." (Flashback? a line up of Big Tobacco executives assures a Congressional hearing that "nicotine is not addictive.")

It was captivating to witness the eminent scientists Drs. Henry Lai, Om Gandhi, Martin Blank and Leif Salford in attendance.

During a discussion between the panel of presenters and the senators about Dr. Lai's seminal work on DNA damage, I wanted to leap up and shout, "Dr. Lai is here. Let's ask him."

At the closing, Senator Harkin commented, "I found this really very interesting and very challenging, and I can assure you we are going to do some follow-up on this."

Sen. Arlen Specter declared, "The issue of children is something we should look at a little more closely … We have a duty to do more by protecting children."

I am grateful to the indomitable Dr. Davis for her skilful and compelling testimony and for organizing the conference that surrounded this hearing. I am hopeful her voice, and the voices of other concerned scientists, will continue to be heard. Loud and clear.

Other US legislators are responding to this message. Christiane Tourtet, BA, Founder of International MCS/EMS Awareness, is bringing this issue to progressive-minded US state governors. Connecticut Governor M. Jodi Rell and Colorado Governor Bill Ritter, Jr. signed proclamations declaring May 2009 as Electromagnetic Sensitivity (EMS) Awareness Month in their states.

In Canada

Dr. Keith Martin, PC, MP, MD – a physician and Member of Parliament – has held public meetings on cell tower concerns and has called for government action:

> Canada needs to engage in an independent assessment of the effects of electro-magnetic radiation on human health, examining the physiological effects of wireless communications on humans. Recent research has created sufficient concerns that we must do this in the interest of the health of our public.

In November 2009 Mayor David Saunders of Colwood, British Columbia, signed the Porto Alegre Resolution, becoming the first city in North America to support this call for the precautionary principle in dealing with wireless technology.

In the UK

The Leader of the UK Green Party, Caroline Lucas, MEP, is also a proponent of dealing with the realities of health hazards:

> Caroline has called consistently for a moratorium on mast installations (cell towers) and the rolling out of new technology until there is sufficient independent evidence to guarantee their safety. **This approach, in keeping with the precautionary principle, is in marked contrast to the position adopted by industry and government** which assumes they are safe until proven otherwise.
> www.carolinelucasmep.org.uk

Another member of parliament in the UK, Dr. Ian Gibson, MP, is supportive. Dr. Gibson's academic training in biology and genetics gives him a special understanding.

In Europe

In 2008 the European Parliament showed it is paying attention, voting by 522 to 16 to urge ministers across Europe to bring in stricter limits for exposure to radiation from mobile and cordless phones, wireless Internet and other devices, stating:

> The limits on exposure to electro-magnetic fields which have been set for the general public are obsolete.

In April 2009, these same European parliamentarians took another step, passing a major resolution that antennas, mobile phone towers and other EMR-emitting devices should be placed a specific distance away from schools, retirement homes and health institutions. The vote was 559 votes in favour, 22 against, with 8 abstentions; the resolution calls for stricter regulation and protection for residents and consumers, especially children.

At a February 2009 conference held in Brussels, Andrzej Rys, MD, Director of Public Health for the European Commission, spoke of the need for continuing progress on new, biologically-based public exposure standards, and implementing precautionary strategies with respect to wireless technologies.

Some European member states have taken action. Wireless Internet has already been banned in public schools in Frankfurt, Germany, and Austria's Medical Association is pressing for a similar ban. Several libraries in Paris have removed their wireless Internet networks due to health concerns. And, in May 2009 it was announced that France is prohibiting cell phone use in elementary schools, also due to the health concerns, after a study on wireless radiation.

In October 2009, the French Senate moved toward banning cell phone use in schools, by children up to about age 14. The amended bill also outlaws cell phone advertising to children. Citizen activist groups are also demanding more cell tower regulation, and are beginning to win some victories both from local governments and courts.

In April 2009, the Mayor of the City of Oullins decreed that new mobile phone masts could not be built within 100 meters of schools or day care centers. What's more, the phone companies are required to make sure the electro-magnetic emissions, within this protected zone, do not exceed the low levels recommended in the BioInitiative Report.

Then a district court cited the Precautionary Principle when it issued an injunction stopping a French telecom company from installing cell phone antennas on a hotel, 15 meters from an apartment building in Paris. The plaintiffs, aged 71 and 83, argued that the new towers would beam radiation directly into their bedroom.

The judgment says, "Even if present scientific knowledge does not enable us to determine precisely the impact of electro-magnetic radiation when it penetrates the communal parts of the building, there exists a risk that cannot be ignored of repercussions on the state of health of the residents who live in it." The court also ordered a 5000 Euro fine against the company for every day its order is not adhered to.

The lawyer for the elderly couple says this decision could set a legal precedent, as it's the first time that the precautionary principle has been used in France as a protective measure for adults.

Another encouraging example: Salzburg, Austria, with the guidance of renowned physician and EMR researcher, Dr. Gerd Oberfeld, has set wireless radiation limits for its city based on biological effects.

And the Liechtenstein Parliament has confirmed its intention to restrict exposure from mobile telecommunication base stations to the BioInitiative Report's recommended levels, by 2013, in sensitive areas such as homes, workplaces, schools, hospitals and public buildings. The companies are threatening to pull out of the country, saying the new rules will make it impossible for them to be financially viable. The government has responded by exploring the possibility of a state-owned telecommunications network.

And health agencies in six nations – Switzerland, Germany, Israel, France, the United Kingdom, and Finland – have issued warnings to limit cell phone use, particularly by children.

Scientists are taking action

As you may know, scientists prefer to take an unbiased, discovery science approach to their research and present their papers within the scientific community. In this case, the recognition that there is growing evidence of harm has brought many of them to speak out publicly.

Elizabeth Kelley, MA, is concerned, as a parent and public health policy expert, about the growth and broad deployment of new technologies that use EMF to transmit wireless signals, and electricity, without independent reviews of the scientific evidence on electro-magnetic fields, or environmental assessments to ensure these technologies will do no harm. She directs the International Commission for Electromagnetic Safety (ICEMS), composed of many concerned scientists whose research findings indicate that industry driven exposure standards are not protective. These scientists are calling for nations to adopt biologically-based standards and a sustained independent EMF research program.

And citizens are taking action

 In the US

The President's Cancer Panel has accepted input from Eileen O'Connor, Founder of the UK Radiation Research Trust. The Panel's report is due in January 2010.

To advance sound public health policy on this ever-increasing public exposure, The EMR Policy Institute – led by Janet Newton and Diana Warren – focuses its activities on the recommendations to protect human health set forth in the BioInitiative Working Group Report.

The Environmental Working Group is a nonprofit organization based in Washington, DC. Their mission includes protecting children, babies, and infants in the womb from health problems attributed to toxic contaminants, including electro-pollution, as well as focusing on federal policies.

Moms for Safe Wireless is a national non-profit group, founded by Christine Hoch, that is dedicated to enhancing the public's understanding of the safety concerns about wireless products.

Another effective voice is the dedicated Camilla Rees, the founder of Electromagnetic Health.Org and co-author with Dr. Havas of *Public Health SOS: The Shadow Side of the Wireless Revolution*.

Here is an inspiring example of what one concerned grandmother can do: when her two grandsons became ill, and she learned that EMR exposure was a factor, Joanne Mueller became determined to raise public awareness. The name of her organization is Guinea Pigs "R" Us.

 In Canada

Martin Weatherall and the Canadian Initiative to Stop Wireless, Electric, and Electromagnetic Pollution (WEEP) have been educating the public, and providing help for electro-sensitive people for many years.

The Health Action Network (HANS) is another group working hard to alert people to this health issue.

 In the UK

Alasdair Philips (Powerwatch), Eileen O'Connor (Radiation Research Trust), Sarah Dacre (ES-UK), and others, are educating the public and government leaders. One of their aims is funding for related research and for safety regulations which are in line with existing scientific evidence. Sarah Dacre is a trustee of ES-UK. This registered charity advises and educates those affected, and the wider population, about electro-sensitivity, its possible triggers and ways to lessen symptoms.

WiredChild is a charity run by a group of concerned parents raising awareness of the potential risks to children of exposure to radiation from mobile phones and other wireless technology.

Mast Sanity is a national group of community members who campaign against the unsafe placement of mobile communications towers.

 In Europe

HESE (Human Ecological Social Economical Project) is an international group of researchers and advocates, many of whom have suffered health effects from EMR.

 In France

PRIARTéM (Pour une réglementation des Implantations d'Antennes Relais de Téléphonie Mobile) is a citizens' group created in 2000. It was the first national organization in France to bring this issue to public view. With the rapid proliferation of mobile communications in France – 50 million cell phone users and nearly 100,000 antennas – the group's advocacy activities now include all wireless communications. With members all over the country, and medical advisors including Dr. Daniel Oberhausen, this organization helps citizens in affected communities fight for quality of life and protection of health.

Next-up is another France-based non-governmental organization set up to protect the natural environment and human health from EMR. Its website has a huge database, offered in several languages, of news reports, videos, government and court decisions and public petitions.

In Sweden

Sweden is one of several countries with support for people with electro-hypersensitivity and several of the world's leading researchers are Swedish. In Gothenburg, Anne-Li Karlsson and Jennie Öberg created a group, along with four others. They have built a radiation-blocking Faraday cage, in the shape of a traditional Swedish country house, to get people thinking about all the sources of radiation in and around their homes. What a great idea.

 In Australia

EMFacts Consultancy, founded in 1994 by Don Maisch, has produced a wide range of reports and papers dealing with various health issues related to human EMR exposure.

The EMR Association of Australia is a consumer advocacy group which is working to alert people about the growing concerns in that country.

Look in the Resources section at the end of the book for the websites of the groups mentioned above.

These are just some of the citizens' groups who are taking action. We will tell you about more groups, in each region of the world, as we learn about them and will post this information on our website.

Waking Up

If you're still reading along here with me, I imagine you're getting a good sense of why these people are so concerned. And they are making progress. I realize this first section throws a lot at you. Please don't put this book down now! There is also progress in the evidence, and the interventions – the many things you can do.

With any wake-up call, including the morning alarm (not, I hope, from your cell phone), it is tempting to roll over and go back to sleep. With our already overwhelming to-do lists, who is eager to take on one more thing?

Some days, I wish I could walk away from this myself; then I see the next child pressing the hot pink cell phone against her head … the baby in the arms of someone on a cordless phone …

As a parent myself, I know there are many others who act quickly when they've seen some sound evidence.

One dedicated mom I know got the message about wireless networks right away, "That's it, I'm ripping that router out of our house and going back to our Ethernet connection."

I did point out that the risk is not yet a scientific certainty. She shot back, "With my children, I don't wait for that."

A doting grandmother also woke up on the spot, "My goodness, I had no idea. My son put this wireless thing in my house so he could get his email in the living room. But he is not here that often. What about their home and my grandchildren? Their rooms are full of electrical things. Please tell me where my son can get this information; I'm calling him right now."

Another parent, whose child has autism symptoms, spontaneously hugged me when I told her that many of these children were thriving in a low-EMR environment, and about a new study with promising results, which we will discuss later on.

And we must not forget that there are many health-enhancing electro-magnetic fields. This is also known as life force, chi – the healing energy in us, and around us.

We generate our own bio-electrical field, which is inherently in harmony with the energies that surround us in the natural world. Another good reason to push back from the computer and step outside more often.

Acupuncturists, and many other practitioners of energy medicine, work with our inner energetic flow to encourage our innate self-healing mechanisms.

In later chapters, we will look at these natural energies and the healing benefits of some of these traditions. The key, as in many things, is balance. We can withstand a certain amount of EMR and other toxicity as long as we are not inundated to such a degree that our innate wellbeing is overwhelmed.

Our basic healthiness, and our ability to heal, are miraculous beyond measure and can be supported in ways we will discuss in this book.

We industrious little beings created this electro-pollution, and we can reverse it, starting with our own homes, where much of this exposure happens.

Every contributor to this book is supporting you, and behind you all the way, to safeguard your family.

We can meet this Radiation Rescue Challenge™. For all of our children.

In the next section we turn to the crucial step of taking stock.

The benefits of this include knowing how much EMR exposure you, and your family, are dealing with on a daily basis; how this may be affecting you, and symptoms and medical conditions that may be related.

And we look at the safer options for your living, working and travelling environments.

This situation is workable. There are solutions.

Difficulty sleeping, headaches, increased stress, trouble concentrating, dizziness, impaired immune systems, food allergies and chemical sensitivities can all be connected with EMR exposure.   Hans Scheiner, MD

Step 2. Know Your Risks

- throughout each day, we are exposed to artificial EMR
- these levels have been shown to have adverse biological effects
- this exposure now begins in utero
- there is a wide range of related symptoms and health conditions
- many medical tests and treatments expose us to EMR
- millions of people around the world are affected by a new condition: electro-sensitivity
- knowing your risks is essential to reducing them

A Family Like Yours?

Before we get to the questionnaire, let me introduce you to a mother, perhaps much like you, or someone you know. I met this woman and her family at our local library.

After discussing books, including this one, she asked if she could read it. It was still in process so I sent her sections to read, inviting her feedback.

Here's an excerpt of what she wrote:

> Sorry that it's taken me so long to get back to you about the second chapter. I think that I had some fear around what I might learn – after all the evidence in part one. So there was some resistance for me to overcome before I could read it properly!
>
> I was definitely intimidated by the checklists in Step 2. I have so many of the symptoms.
>
> I think also, that it might be useful to suggest to the reader that they do a cost/benefit analysis ie. what is the potential long term risk of EMR exposure vs. what is the potential loss that they will incur by restricting their EMR exposure.
>
> They will probably find that many of the lifestyle changes required to restrict EMR are not life-threatening--but are actually life enhancing.

As you'll see in the questionnaire that follows, there is a surprisingly long list of symptoms and conditions that may be linked to EMR exposure. It can be challenging to figure out whether EMR is a factor in any individual's case, though there are ways to measure this and to see whether reducing your exposure reduces your symptoms. There's more about this in Steps 3 and 4.

The Radiation Rescue Questionnaire©

The information contained within the questionnaire does not provide medical advice and is not intended for use in medical diagnosis or treatment. In the case of any disease, you should always consult your health care practitioner.

You'll notice this survey is not scored; that's because there is no way at the moment to weight the effects of the various exposures.

For example, did you know that **Dr. L. Hardell's early research indicated that 500 minutes a month of cell phone use significantly increases brain cancer risk – as much as 140%.**

Some other studies show up to a 300 per cent increase when using the cell phone between 500 and 1000 minutes per month.

Source: Wire Technology Research, Second State of the Science Colloquium Book; January 2001 Journal of the American Medical Association; Early Hardell studies.

No one yet knows how much the risk might be increased if combined with exposure to wireless radiation, portable phones and/or overhead power lines. As Professor Olle Johansson cautions, we are all part of that experiment.

This subjective questionnaire is a work-in-progress. I created this from researching the literature, and from several documents compiled by others.

I had the editing assistance of Dr. Heather McKinney, and Sarah Dacre – an advocate in the UK. My thanks to them. In 2007 Dr. McKinney and I reserved editorial control of this document. Please contact me if you wish to use it in any form.

Section I – Symptoms: Please check those that may affect you when you are exposed to wired, and/or wireless, devices.

- ☐ Abdominal pain
- ☐ Aggressive moods
- ☐ Depressive moods
- ☐ Dry or painful eyes
- ☐ Erratic blood pressure
- ☐ Excessive sweating at night
- ☐ Fatigue
- ☐ Hair loss
- ☐ Headaches
- ☐ Heart palpitations/irregular heartbeat
- ☐ Inability to focus
- ☐ Irritability
- ☐ Libido disturbances
- ☐ Light-headedness/dizziness
- ☐ Loss of appetite
- ☐ Memory loss
- ☐ Menstrual flooding/irregularities
- ☐ Metallic taste in mouth
- ☐ Nausea
- ☐ Nightmares
- ☐ Pain/discomfort in the heart area
- ☐ Pain in the head, neck, shoulders, back
- ☐ Panic attacks
- ☐ Ringing of the ears
- ☐ Sensitivity to noise and/or light
- ☐ Sleep problems
- ☐ Skin rashes/bumps/dryness
- ☐ Tingling and/or numbness
  – in the head, hands and/or feet
- ☐ Vision problems

Section II – Conditions: Please check the items that apply only to your personal health history.

- [ ] Adrenal overload
- [ ] Allergies
- [ ] ALS/Motor Neuron Diseases
- [ ] Alzheimer's Disease
- [ ] Autism Spectrum Disorder
- [ ] Brain aneurism
- [ ] Cancer
    - [ ] Eye
    - [ ] Ear
    - [ ] Brain (adult or child)
    - [ ] Breast
    - [ ] Testicular
    - [ ] Other
- [ ] Leukemia (adult or child)
- [ ] Lymphoma
- [ ] Candidiasis
- [ ] Cataracts
- [ ] Cardiovascular disease
- [ ] Chronic Fatigue Syndrome
- [ ] Cold or flu (persistent)
- [ ] Dementia
- [ ] Fibromyalgia
- [ ] Food sensitivities
- [ ] Heart attack
- [ ] Heavy metal toxicity

- ☐ High blood pressure
- ☐ Infertility
- ☐ Insomnia
- ☐ Irritable Bowel Syndrome
- ☐ Leaky gut syndrome
- ☐ Learning Disorder
    - ☐ ADD
    - ☐ ADHD
- ☐ Lupus
- ☐ Lyme Disease
- ☐ Migraine, or other severe headaches
- ☐ Miscarriage
- ☐ Multiple Chemical Sensitivities (MCS)
- ☐ Multiple-sclerosis
- ☐ Parkinson's Disease
- ☐ Sleep disorder
- ☐ Stroke
- ☐ Systemic infection
- ☐ Thyroid gland disorders
- ☐ TIA (Transient Ischemic Attack)

Do any of these conditions feel worse when you are exposed to wired, and/or wireless, devices?

☐ No   ☐ Yes   ☐ I don't know

Step 2. Know Your Risks

Section III – Sources of EMR Exposure – Wireless Devices:
Please check 'Yes', or 'I don't know' in the columns, as appropriate. Leave an item blank to signify 'No'.

	Yes	I don't know.
Do you regularly use a cell/mobile phone?	☐	
If Yes:		
More than 5 hours daily?	☐	
More than 2500 minutes per month?	☐	
More than 500 minutes per month?	☐	
In your car?	☐	
Prior to 1996?	☐	
Do you experience Symptoms with use?	☐	☐
If No:		
Do you experience any Symptoms around cell/mobile phones?	☐	☐
Do you use a wireless in-your-ear cell phone?	☐	
Do you regularly use a hand-held PDA (personal digital assistant)	☐	
If Yes::		
More than 5 hours daily?	☐	
More than 2500 minutes per month?	☐	
More than 500 minutes per month?	☐	
In your car?	☐	
Do you experience Symptoms with use?	☐	☐
If No:		
Do you experience Symptoms when you are around PDA's?	☐	☐

	Yes	I don't know.
Have you ever checked the SAR (radiation absorption level) of your cell phone/PDA?	☐	

Do you...

	Yes	I don't know.
Have your cell phone and/or PDA powered on at night?	☐	
Use your cell phone and/or PDA as an alarm clock?	☐	
Use a headset or ear piece with your mobile?	☐	

If Yes: Check the type you use

☐ Wireless ☐ Wired

	Yes	I don't know.
Use a wireless game station, or wireless video box?	☐	
Use a wireless device with Internet access, or that downloads music, movies or other wireless transmitted data?	☐	
Use a portable satellite, or wireless broadband, radio?	☐	
Have a GPS, satellite radio, or wireless system, in your car?	☐	
Drive a commercial vehicle with a satellite/GPS locator?	☐	
Use a personal computer connected to wireless Internet?	☐	
Have wireless Internet access....		
a) in your home?	☐	
b) in your workplace, school wireless?	☐	☐
c) in your neighborhood?	☐	☐
d) in your city?	☐	☐

	Yes	I don't know.
Live or work near a cell tower, or mast?	☐	☐
If Yes:		
a) Within 100 metres?(110 yards)	☐	☐
b) Within 200 metres? (220 yards)	☐	☐
Work with, or live near, radar devices/systems?	☐	☐
Use an amateur radio, 2-way or CB radio?	☐	
Have a cordless/portable phone?	☐	
If Yes:		
a) In your home/office/school?	☐	
b) In your bedroom?	☐	

c) Your total number of cordless phones, and/or baby monitors is:

1____ 2____ 3____ 4____

Are your cordless/portable phones	Yes	I don't know.
900 MHz	☐	☐
2.4 GHz	☐	☐
5.8 GHz	☐	☐
DECT 6.0	☐	☐
Around wireless "hot spots", or devices, do you experience Symptoms from Section I?	☐	☐

Other exposures:

	Yes	I don't know.
Are you an airplane pilot or flight attendant?	☐	
Do you travel often, through security body scanners?	☐	
If yes, do you experience any Symptoms with exposure?	☐	☐
Do you work at supermarket checkouts/libraries near scanners?	☐	
If yes, do you experience any Symptoms with exposure?	☐	☐
Have you had a CT scan?	☐	
If yes, did you experience any Symptoms with exposure?	☐	☐
Have you had an MRI?	☐	
If yes, did you experience any Symptoms with exposure?	☐	☐
Have you had medical and/or dental X-rays?	☐	
If yes, did you experience any Symptoms with exposure?	☐	☐
Have you had a long-term hospital stay – more than one week?	☐	
If yes, did you experience any Symptoms with exposure?	☐	☐

Section IV – Sources of EMR Exposure – Electric Devices:

Please check 'Yes', or 'I don't know', as appropriate. Leave an item blank to signify 'No'.

	Yes	I don't know.
Do you...		
Use an electric blanket, and/or heating pad?	☐	
Sleep on a/an:		
a) Electric adjustable bed?	☐	
b) Metal bed frame?	☐	
c) Coiled mattress/box springs?	☐	
d) Electrically-heated water bed?	☐	
Sleep within 2 metres (2.2 yards) of electric devices, including: a clock, radio, compact fluorescent, or low voltage halogen, lights?	☐	
Sleep within 6 metres (6.5 yards) of an electrical fuse panel?	☐	
Stay in a hotel more than five nights per month?	☐	
Regularly use a hairdryer and/or electric shaver?	☐	
Use a microwave oven?	☐	
Are you often by the front burners of an electric stove while they are operational?	☐	
Are you often near electric heaters or "off-peak" or "overnight" electric storage heaters?	☐	
Do you...		
Live/work/school near overhead power lines?	☐	☐
Live in a densely populated urban area?	☐	
Work/live near electrical transformers?	☐	☐
Work/live near a electrical sub-station?	☐	☐

	Yes	I don't know.
Live/work near an airport?	☐	☐
If Yes:		
a) Within 0-5 km? (0-3 Miles)	☐	☐
b) Within 5-15 km? (3-9 miles)	☐	☐
Work/live in a brightly lit room more than 5 hours daily?	☐	
Work with power tools?	☐	
Work with other electrical, or high frequency, equipment?	☐	
Does your home/work have dimmer switches on any lights?	☐	
Do you have low voltage halogen, tube or compact fluorescent, lights at work and/or at home or school?	☐	
Do you live/work in an area with high radon gas?	☐	☐
Do you drive/ride in a hybrid car (gas/electric), or a car with wireless Internet and other high-tech electronics?	☐	
Do you experience Symptoms from Section I around electric devices?	☐	☐

As you know, when artificial electricity was introduced it was a whole new kind of energy that created many obvious advantages and some great gadgets. Obviously, there's harm if you put your finger in a live socket, but generally, we have come to view electronics as benign; this may explain why many people sleep with a cell phone close by their heads without concern. I'm hoping that by now you are viewing our electrified conveniences in a more cautious light.

You may well be thinking, "Okay, I've got it. What can I do about this problem?" So, let's get on with it.

To begin, we need a quick crash course.

Why do we need to know the EMR basics?

You don't really need to be an expert in the various kinds of radiation to reduce your exposure. However, you want to be sure that you are decreasing, not increasing, your electro-magnetic load. (This is an important reminder for the get-on-with-it types who dive into assembling something without reading the instructions.)

Here's an example of a little knowledge being a less-than-safe thing... I visited a friend who has a few testing meters. I was surprised to see she was standing close to her microwave oven when it was on. I asked gently if she felt this was safe. She said, "It's okay, no problem. I tested it and it's fine at this distance." She was using her Gauss meter, which tests the magnetic field around this plugged in appliance. She didn't, however, test the microwave radiation with the instrument which measures that – the RF meter. Not only did she test the wrong exposure, she missed the most harmful one. More on this in a few pages…

The Electro-magnetic Spectrum

So here's what you need to know.

An electric field is the energy surrounding wiring in the walls, or an electric device, extension/power cord that is plugged in. (Note: there is an electric field around everything that is plugged in, even when it is turned OFF. Surprised?)

A magnetic field is the energy surrounding the same wires or appliances when an electrical current is flowing – for example, the lamp is turned on.

Radio frequency (RF) radiation – these wireless waves (fields) are also called microwave radiation. Here the frequency is so high that small packets of energy are released from the antenna – as in mobile devices. In the case of a microwave oven, you can cook with these waves.

Electro-magnetic radiation (EMR) – also known as **electro-magnetic fields (EMF)** – is the over arching term for most of the spectrum – see below. Technically, EMR/EMF includes everything but the highest frequencies (ionizing radiation – X-rays etc).

The Three Windows

A. Lower Frequencies Electricity	B. Higher Frequencies Microwaves/Radio Frequency Wireless except microwave ovens	C. Highest Frequencies
Power lines		Nuclear material
Transformers	Microwave ovens	The Sun
Electrical wiring	Fluorescent lights	Cosmic Rays
Plug-in appliances	Satellite uplinks	Gamma Rays
Electric blankets	Cell tower antennas	CT Scans
Electric heaters	Wi-Fi, wLAN, WiMAX	Medical X-rays
Electric shavers	Cell phones - PDAs	Dental X-rays
Hair dryers	Cordless phones - DECT	Airport security scanners
Power tools	Wireless: devices/games/speakers	Ultraviolet (fluorescent lights)
Office equipment	headsets, earpieces	TEST WITH instruments that
Lamps, light fixtures	RF identification tags	measure X and gamma rays,
Dimmer switches	Baby monitors	like a GEIGER COUNTER
Electric clocks	Security systems	or a SCINTILLATION
Power cords	Dimmer switches	COUNTER.
Extension cords	Police and other radar devices	
Power bars	Pagers, Walkie-talkies	
Computers	Radio frequency sealers	
TVs - Video Display Terminals	Radio AM/FM, CB radios	
Plumbing current	GPS systems	
Electric Fields - TEST WITH ELECTRIC FIELD METER Magnetic Fields - TEST WITH GAUSS METER	TEST WITH RF METER	

As you see, there are three basic kinds of frequencies.

A. Lower Frequencies: (electric and magnetic)

Electric fields are emitted by:

- Household electrical wiring in ceilings, walls and floors; wiring that powers lights and appliances
- Everyday power cords, extension cords, and power bars when plugged into an outlet, even when the appliance is not turned on
- Electric blankets, waterbeds, TVs, computers and all other electronic devices
- All electrical appliances, lamps and lighting

Step 2. Know Your Risks

- High voltage power lines

Electric fields are tested by:

- An electric field meter

Magnetic fields are emitted by:

- High voltage power lines
- Electrical wiring errors and current on the electrical grounding system, water and gas pipes
- Electric meter and fuse box, dimmer switches, all power transformers/AC adapters
- Fluorescent and low voltage halogen lighting, alarm clocks
- Heating elements in stoves and ovens when heating
- All appliances (refrigerator, washing machine, furnace, hair dryers, exhaust fans etc…)
- All CRT (cathode ray tube) VDT (video display terminal) televisions and computer monitors (in some cases, LCD displays are a better choice but some still emit hazardous levels of EMR.)
- Automobile engines

Magnetic fields are tested by:

- A magnetic field, or Gauss, meter.

B. Higher Frequencies: (the middle – the wireless – window) Microwaves/Radio Frequency fields are emitted by:

- Baby monitors, wireless security systems, wireless game pads, and microwave ovens
- Cellular tower base stations, cellular phones, hand-held PDAs,

all cordless phones, wireless technology
- Wireless computer networks and Internet access routers
- All other wireless communication devices
- Some other sources include: fluorescent lights, PUC (under counter lights), some remote control toys, laptops in wireless mode, and some hydro, gas and water meters

Higher Frequencies are tested by:
- An RF meter.

C. The Highest Frequencies:

The third kind of energy shown on the chart is a powerful force, which can create significant cell damage if there's enough power, proximity and duration. This is why it is now recommended to limit X-ray exposure to medically-urgent situations. Other sources include: radioactive mineral deposits, radon gas, nuclear facilities and fallout, depleted uranium such as some troops deploy in combat, and household smoke detectors that contain americium dioxide.

We are all exposed to cosmic radiation to some degree; it's most significant at high altitudes, or in airplanes. At 36,000 ft. the levels can be hundreds of times higher than on the ground.

The Highest Frequencies are tested by:
- Instruments that measure X and gamma rays, like a Geiger counter or a scintillation counter.

Whether or not these or any other exposures are harmful depends on some significant variables. With each exposure you want to consider these factors.

Six Key Factors:

Evidence of harm – as we have discussed, even electro-magnetic levels far below existing 'safe' standards, not high enough to heat human tissue, cause adverse biological effects.

The evidence? See the first section of this book, or some of the Resources' websites. The Resources are at the end of the book.

Intensity – the power of the specific source: the type and power of your cordless phone – analog or digital – 900 MHz to DECT 6.0; and the range of the cell towers, broadcast antennas, power lines, wireless networks etc.

Proximity – how close you are to the source. Is the cell phone against your ear; the wireless laptop on your lap; the cordless phone base nearby; the cell tower, transformer, electric sub-station, or power lines near your house/school/workplace? Your neighbors' wireless network is exposing your family, but is less harmful than having one in your own home – due to proximity.

Duration/frequency – if you have a high-powered electric heater (intensity) on for ten minutes to take the chill off the room, no big deal. If you sleep beside it, or have it on for hours every day (proximity, duration and frequency) that's more of a concern.

Vulnerability – fetuses, infants, children or teens, with thinner skulls, still-growing brains and more aqueous brain tissue, are at greatest risk. Older people, and those with compromised immune systems, learning disorders and other medical and neurological conditions, are also vulnerable.

Combination of factors – the more of these factors that are present, the greater the danger, as you would presume. A baby sleeping near a wireless monitor, an electric outlet, and a cordless phone, in a home with a wireless security system and wireless Internet is surrounded by multiple, accumulating exposures.

Before I knew about all this, I gave my 85-year-old mother a heating pad to improve her circulation. She began sleeping with it held against her tummy. Shortly afterwards she began to experience uterine bleeding and was taken to hospital to be checked out.

Fortunately, this was about the time I found out about EMR, and that it could cause internal bleeding. My mother's doctor dismissed quickly this information, but we removed the heating pad anyway, and the bleeding stopped within 24 hours. (As with our daughter's metal canopy bed frame, I couldn't bring myself to give the heating pad to anyone else.)

All factors were present: the EMR intensity of the heating pad, the proximity of having it next to her body, the duration and frequency of using it all night every night, and her vulnerability from age and health conditions.

Another example: imagine a fragile premature infant in an incubator, hooked up to all kinds of life-saving electrodes and monitoring equipment.

Adding to the infant's exposure is the hospital's electro-magnetic load – fluorescent lighting, wireless systems etc. The hospital may even have a large cell antenna on its roof.

Before we go to Step 3 – Seek Safer Solutions, let's check in again with our library mum:

Here's a few family changes since reading part of your book:

Read Step 2 last night. Thought about how I use my new multi-task cell phone to listen to music and how we also use it for an alarm. Went and turned off my cell phone for the night, as well as both computers which were humming away. Put 'get battery alarm clock' on my 'to do' list.

Am contemplating getting rid of my new all-in-one phone and scaling back to the old cell phone.

Wondering if it is possible to run the business without a cell phone.

As you know from meeting me at the library, we read a lot. Feel even better about that now knowing why it's good to have time away from electronics.

Am feeling like we need to get outside more.

Step 3. Seek Safer Solutions

Here's what you need to know about the EMR-emitting devices in your home, office, school ...

In this section we look at the specific devices – cell phones, headsets, cordless phones, baby monitors, Internet access etc. – and hear from some of world's leading technical experts including: Alasdair Philips, Rob Metzinger, Larry Gust and Stan Hartman, Katharina Gustavs and Chris Anderson. This is what we, the consumers, have needed to know.

This is probably more than you *want* to know about many of your favorite things. I'm hoping that you won't go from techno-crazed to techno-phobic – there is a healthy balance in the middle way. The technical truth about these devices has been a three year discovery for me; I have had time to go through my stages of denial, anger, action and adjustment.

I don't want you to feel so swamped with these details that you abandon this rescue project altogether. If you find yourself feeling exasperated – "Oh no, not my hairdryer" – please take a break and come back to this another day.

Over time, if you can just tackle a few of the 'biggies' – such as the cordless phone – you have done a lot to safeguard your family.

How can you decide whether devices are safe or not?

Let's use the analogy of sun tanning. You know that while we need some exposure to the sun (evidence of benefit), too much can be hazardous (evidence of harm). It follows that caution is advised when being outside in the mid-day sun (intensity), or using sun lamps or tanning beds (intensity and proximity).

According to our experts, if you are using a powerful device (intensity), on a regular basis (duration), wearing it on your body, sleeping near it, or holding it in your lap – or against your head (proximity), you are at risk. You can see why there's so much concern about cordless and mobile phones.

Mobile Alert: Even Higher-Powered Tech Toys on the Electronic Horizon

An eye of awareness can wake us up when we see advertisements touting the latest high-tech toys: a compact wireless "netbook", a formidable DECT 6.0 cordless phone station to control all the functions in your home, a pink pendant cell phone that nestles in your cleavage, a wireless exercise pad for "Wi-Fi fitness", and mobile online banking services – right from your cell phone.

Here's one to jolt your alert button: we are promised a 4G multi-function mobile that will "be like your very own butler in the palm of your hand. It'll do everything but tuck you in at night."

Get your radiation-testing meter ready.

The Key Question: Does it work without wires?

Many people are concerned about living near power lines, or electric sub-stations. This risk is validated by extensive research. However, we are just learning of the risk of the radiation from cell towers.

Somehow these massive structures do seem menacing (of course, that's why they are often concealed in metal church crosses, or disguised as palm trees). Yet, the digital cordless phone sitting on your kitchen counter looks so benign. The new digital baby monitor seems to be essential for safety-minded parents. And in this digital age it seems an educational advantage to start our children young in becoming at home with computers – usually with wireless connection to the Internet.

For something to communicate without wires, it uses some form of radiation to send and/or receive data through the air. Most of these waves travel easily through cement walls. This means they are also going through the skull, and the rest of the body.

The Key to the Safer Solutions

Stan Hartman, an environmental health consultant, offers us key advice:

> The most important thing for people to keep in mind when trying to reduce electro-smog is, first of all, to replace all wireless technology (except infrared remote controls for TVs and other things) with wired alternatives.
>
> Get rid of all bluetooth devices, and replace other wireless gear like cordless speakers, headphones, keyboards, mice, etc.

If the data transmission is going through a hard-wired cable, a plugged-in phone line, or a fiber optics system, then it does not present that kind of exposure. These differences form the basis of the safer solutions.

When you have this awareness these decisions seem to arise instinctively. And wiser choices become obvious. You can decide, for example, how to set up your online banking.

If you're using a device that has a thin LCD screen (vs. the larger cathode ray terminal) and hard-wired Internet access (instead of wireless), you can enjoy that convenience with much less risk than using your cell phone or PDA.

The Sources and Recommendations

In this section we'll look at exposures from various sources, explain the experts' concerns about them, and suggest safer solutions.

It is important to know the difference between making something safer, and making it safe. For example, when you add a filter to a plain cigarette does it make it safe? If it is a low tar cigarette does it make it safe? If you only breathe in second hand smoke, does it make it safe?

As you read through this section, you will get an idea of the exposures you want to reduce. At the end of this Step there is a preliminary Action Plan to help you prioritize; you'll find a more detailed version at the end of Step 4.

We are going to start with the devices from the higher frequencies – the middle of the EMR Spectrum.

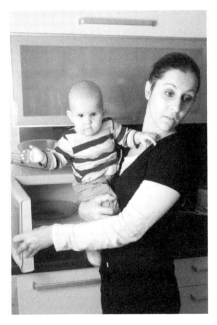

Microwave Ovens

The Concern: Larry Gust, an electrical engineer, tells us why that handy appliance is not as harmless as it looks.

(LG) "**All microwave ovens leak RF radiation (radio frequency, microwaves) when they are in use.** This radiation covers a surprisingly wide area – as much as 3 to 30 feet, and in older models up to 60 feet. Most people think that microwave ovens do not have any negative health consequences. After all, if they did the government would not allow this product on the market. Think again.

The Soviet Union conducted extensive microwave research after finding German microwave ovens and research data at the end of WWII. Unhampered by capitalistic trade associations, Russian research has been more extensive than that in the US. This article summarizes early Russian and other more recent research.

Microwave ovens use electromagnetic energy that vibrates 2.4 billion times per second. This energy acts on the molecules in food, particularly water molecules, causing them to vibrate rapidly. This rapid movement generates friction and thus heat. Vibration is so violent that molecules are often torn apart or distorted, thereby changing the chemical makeup of the food. Additionally, this appliance leaks microwaves into the environment around it.

Studies of the thousands of Russian workers exposed to radar microwaves in the 1950's showed health effects so severe that the Russian microwave oven leakage limit was set at 1,000 times less than the US government leakage limit.

In 1976, microwave ovens were banned in the Soviet Union after researchers discovered that these devices caused a breakdown in the human (and animal) energy field. They also found degeneration of cell membranes and electrical nerve impulses in the brain, disruption of hormone production, and disturbance in various brain waves, leading to memory and concentration loss, slowing of intellectual process and interrupted sleep. Microwave ovens were allowed after 1987 when, under Perestroika, business interests pressured Gorbachev into changing the regulations that did not fit in with free-trade practices in the West."

Microwaving baby milk is not recommended. In the December 9, 1989 issue of the British medical journal, *Lancet*, Dr. Lita Lee reported that heating formula in a microwave oven converted some of the amino acids in it into synthetic forms that are not biologically active. One was converted to a toxic form.

Dr. Lee is concerned:

> It's bad enough that many babies are not nursed, but now they are given fake milk (baby formula) made even more toxic via microwaving.

Microwave Ovens – Safer Solutions:

Rob Metzinger (RM): "Replace microwave ovens with safer appliances and methods."

- If you absolutely must have a microwave oven get a smaller, lowered powered one and stand at least 15 feet away – even farther, to be on the safe side – when it is in use.

- Avoid heating baby formula in a microwave, and keep infants and children far away when it is on.

- Better alternatives: keep a small pot or pan with a lid for quick heating/reheating of food and liquids. Avoid non-stick coated pans. (As soon as they are the least bit scratched, you can ingest the toxic little flakes.) At work, heating your lunchtime leftovers in a toaster oven, or on a hot plate with the small pan and lid, takes a couple of minutes longer, but is a safer solution, the experts advise us.

- If you need something defrosted, take it out of the freezer a night or two beforehand and defrost in the fridge.

Cordless Phones

The Concern: You may have known about the hazards of microwave ovens, and been wary of cell phones, but did you know about the health hazards of cordless phones? These conveniences emit high levels of radiation, particularly the newer ones, even when not in use.

Did you know that multiple cordless phone sets that come with a base station, and a charger for each handset, emit two kinds of radiation? The charger sends out harmful magnetic fields within 1 to 3 ft. (The emission of magnetic fields applies to all phones, cordless or not, that come with a cube-style adapter.)

The main base station of a cordless phone, however, is the more dangerous, as the antenna radiates microwaves continually for hundreds of feet/meters. Microwave radiation is also coming from the handset antenna. This is why all the experts urge us not to allow young children near cordless phones. Replacing them should be a key factor in your Action Plan.

From a January 2006 report by the German Federal Agency For Radiation Protection, *The Radiation Source at Home – cordless phones radiate unnecessarily:*

> A cordless phone (of DECT standard – 5.8 GHz or DECT 6.0) is often the strongest source of high frequency electromagnetic radiation in a private home. To renounce your cordless phone as a precautionary measure will contribute to minimizing your family's radiation exposure.

DECT TECHNOLOGY

Beginning in 2008, phones of the DECT 6.0 generation started flooding the US market. DECT is the abbreviation for Digital Enhanced Cordless Telecommunication technology. In the last decade, only Europeans used this technology and were confronted with its headaches.

The 2.4 GHz band has been used by the majority of cordless phones in the last five years. However, this band has become crowded with devices such as microwave ovens, wireless area networks (WLAN), Bluetooth, cordless phones and remote control devices. This has led to interference and performance issues so the manufacturers moved to the 5.8 GHz frequency band and then to DECT 6.0 technology.

The problem with the DECT phones is that the base station is permanently sending (radiating) 24 hours a day, no matter if the system is in use or not. Indoor environmental consultants have rallied against their use for decades. There were so many outcries that it had an impact. Major European manufacturers now offer environmentally safer versions, with sleep mode or reduced output in standby mode.

Case Study

Clients are concerned about the installation of a new cellular base station to be erected on the roof of a recreation center across the street from their house and asked to have their home assessed prior to the installation.

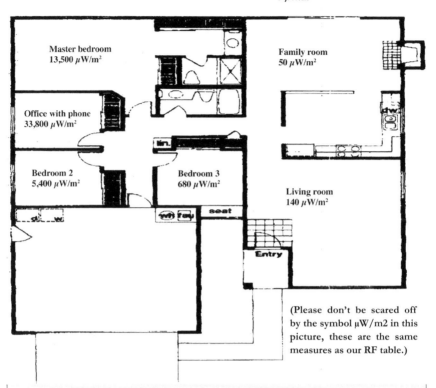

(Please don't be scared off by the symbol μW/m2 in this picture, these are the same measures as our RF table.)

RF Exposure from One 2.4 GHz Cordless Phone in a House (these levels are *only* from the cordless phone)

When Peter Sierck sent me this paper on cordless phones, with the RF levels you see in this graphic, I couldn't quite believe that these readings were coming from *one* cordless phone.

He assured me that, "yes, these RF levels are from one 2.4 GHz cordless phone in the house, no wireless, cordless or other RF sources."

Can you just imagine the levels with a few 5.8 or DECT 6.0 phones in a similar house equipped with wireless Internet?

Another widely respected technical expert, Alasdair Philips, the director of Powerwatch in the UK, cautions:

> It's likely everyone in a house with a (DECT) digital cordless phone will be constantly exposed to levels of microwave radiation far higher than safe levels.

Dr. Hans Scheiner, the physician in Munich, also warns us about digital phones (virtually all new/recent cordless phones):

> Digital signals are more aggressive than analog. Both are harmful. However, digital signals are more aggressive, and more harmful, because they pulse and have peaks. This is a problem as the body's cells react negatively to the sharp peaks.

I know; all these health hazard alerts about cordless phones were a shock to me too. I invite you to consider this as good news and bypass the stage of remorse at having placed these phones throughout the house, and choose instead to hear this message.

Ready for another story? A good news one? I have a dear friend who moved into a small duplex near the waterfront, several years ago. She was thrilled to have found this 'fixer-upper' with a wide porch where she could sit and watch the waves; until she discovered that those were not the only waves hitting her haven by the sea. Since moving in, for no known reason, she was not sleeping well and her Chronic Fatigue symptoms were worsening daily.

Right across from her, not too far from shore, was a small island sprouting several broadcast antennas. This was around the time of the dawning of my wireless wake up call so I shared with her what I was learning. She is a doctor with a keen and curious mind so she delved head first into the data. And soon took action.

My friend hired an EMR assessment consultant to do some testing. He identified what we had both suspected – radio frequency radiation from the antennas across the water, and the wireless network in her home. She replaced her wireless with an Ethernet cable, and that seemed to help, but not enough. The consultant had also discovered another source that surprised us, but not him.

The person living in the other half of their duplex had a digital cordless phone that was blitzing the whole building, and way past the surrounding garden. (This may seem more credible to you after just seeing Peter Sierck's diagram of the house plan.)

Fortunately, the elderly woman who lived there was open to hearing about this, although she couldn't quite believe it, "My dear, I have had a cordless phone for years. I keep it beside my chair, and my bed, as I am slow getting up."

After she replaced it with a landline – with a sufficiently long cord – she and my friend both reported sleeping much better, and a reduction in several previously unexplained symptoms.

What similar benefits might there be for you and your family? Read on ...

Cordless Phone Models:

There is a huge difference in radiation exposure from lowest to highest power. **Note: One gigahertz (GHz) is equal to 1,000 megahertz (MHz).**

The 900 MHz analog (if you can find one) emits much less radiation than the newest DECT 6.0 model. Evidently, you can walk three blocks with this new mega-powered unit.

This list goes from the least to most harmful according to our technical experts:

900 MHz analog – hard to find, but a good option (only emit RF when in use)

900 MHz digital – avoid the DSS (digital spread spectrum)

2.4 GHz analog – higher levels but the analog signal is less of a concern than the digital

DECT: 2.4 GHz, 5.8 GHz, and 6.0 DECT (1.9 GHz) – all use the same pulsing digital technology and the base station emits RF all the time. Radiation levels vary depending by model and manufacturer. The new DECT 6.0 models can emit extremely high levels – similar to the exposure from cell phones.

The environmental consultant, Katharina Gustavs, advises us that, "some of the DECT cordless phones operate at a higher frequency, which is not detected by all RF meters. In this case, the meter may show nothing, or give a false low level."

Children and Cordless Phones:

Many researchers warn us that children are more vulnerable to the effects of radiation. The child's small size will encourage whole-body exposure, and rapidly dividing cells that are part of the child's growth processes are highly vulnerable to radiation damage.

Do not let a child use a cordless phone; never put the base near children's sleeping or playing area, or use one near them.

Accepting the experts' advice that cordless phones are harmful is most challenging, do you agree? We lived with them for ages. But not with this digital signal and powerful intensity.

Remember, constantly amped up electronics are not being tested for biological effects. And, there are slews of hyper-smart techies firing on all cylinders to flood the market with wireless gadgets. Unsupervised. Try testing the EMR levels in an electronics store.

Cordless Phones – Safer Solutions:

(RM) "Replace all DECT/digital cordless phones with landline telephones."

Every one of our technical experts, as well as the other contributors, advise replacing cordless phones. This might seem daunting? Tiresome? But as you learn more it will seem like an essential step in your radiation rescue and worth the hassle of stepping over a phone cord now and then.

Radiation Rescue – The Essentials
Corded Landline Phone

- No microwave radiation

- Replacing cordless phones with landlines may be the best reduction you can make in your family's EMR exposure

- Battery-operated ones have minimal electric field and work in power outages

Every expert I spoke with agrees on this. And if you test cordless phones for microwave emissions (yes, the same kind of radiation) you will understand, and dust off your old corded, landline phone, or go out and buy one.

If you want call display, you need a power source; best option is the battery-operated one, you don't even need to plug it into the wall. You can also get good ones with speakerphone function, so you can still multi-task.

An extended coiled cord on the handset enables you to wander even farther while talking on the corded phone. No, it's not as handy as a cordless phone, but there's no comparison in terms of radiation risk.

If having a cordless phone is non-negotiable for you at this time, consider choosing a 900 MHz analog model – if you can still find one.

The advantage is that it doesn't radiate 24/7, just when you are on a call. So it's recommended to keep those discussions brief, "I'll call you back on the landline .."

I went to see if you can still buy the lower-tech 900 MHz and the salesperson looked at me as though I had asked for two soup cans and a string.

Standing beside a stack of bright blue and black boxes of the hottest DECT 6.0, she declared, "Oh, I don't think they make those anymore; no one wants them. People want the latest, the most high-powered."

Cell Phones, PDAs

The Benefit:

In emergencies mobile phones can be a life-saver, literally, and are unlikely to be harmful if use is very brief, limited to urgent calls, and the phone is kept powered off at all other times.

The Concern:

Near Field and Far Field – another bit of crucial information

Cell phones/PDAs emit a laser-like beam of radiation from their antennas, which may be sticking out or embedded in the device. The near field radiates about a foot from the antenna; the most intense plume is within a 7-inch diameter circle around the antenna.

This powerful blast of near field radiation goes into your ear canal and directly to your brain. It also affects your eyes.

The far field, also radiation that causes concern, goes through everything in its path.

The reason you hear that a headset is recommended is that it removes the near field exposure from the area closest to your head and brain. While in a relative sense, it's safer to keep the antenna's near field as far from your brain as possible, as Rob Metzinger explains, the conventional wired and wireless headsets can work against us by increasing the exposure to other sensitive tissues. (There's more about headsets and earpiece phones later.)

Whether or not you're using a headset, when your mobile is clipped to your belt, or carried in your pocket or shoulder bag, you are radiating breasts (remember the pink pendant cell phone?) testicles and other sensitive body parts including organs like liver, kidney and ovaries. There are also concerns about the effect on the blood-producing bone marrow in your hip.

The intensity of radiation lessens with distance, but even the far field is not considered safe, as the information-carrying radio waves are still pulsing through your body for quite a distance.

The more powerful the device, the more intense the near and far field exposures, as you would imagine. Before you grab the latest multi-function mobile, consider testing it with a RF meter. Make sure the phone is transmitting when you test it.

(KG) "The concern with cell phones is not only the very high levels of RF radiation but they also generate very high pulsing magnetic fields, even in standby."

Pregnant Women, Children and Cell Phones

The Concern:

Ronald Herberman, MD, is the first director of the University of Pittsburgh Cancer Institute, a National Cancer Institute (NCI) designated Comprehensive Cancer Center. Dr. Herberman, and his international panel of experts, caution:

> The developing organs of a fetus or child are the most likely to be sensitive to any possible effects of exposure to electromagnetic fields. Do not allow children to use a cell phone, except for emergencies.

Dr. Hyman, like me, is a concerned parent and advises:

> Parents need to know that our children's developing brains, immune and endocrine systems between the age of conception and mid-twenties (particularly the brain and nervous system) are developing and more vulnerable than an adult's to many toxins, including EMFs. Having a cell phone does not necessarily make your child safer.

There seems to be a consensus among experts that children should not use cell phones, as exposure at a young age is associated with many health problems. Let's hope more governments will follow the example of France.

Alasdair Philips (AP): "Microwaves from mobile phones penetrate the head, and the younger you are, the further into the brain they go, as the skull is thinner and the brain smaller.

Also keep children away from anyone using a mobile phone and don't have one on standby anywhere near a child – on the stroller etc."

Children should not hold any wireless device on their laps, particularly when it is online accessing the Internet, sending or receiving – text messaging etc."

With all of these experts' warnings are you having a hard time taking this in? I know, just about every home is replete with wireless devices. Our children are using them at school, we give them as gifts, they are everywhere.

Another mom who was reading this book commented:

> When I walked into our children's bedrooms, after reading your book, I suddenly realized how many electronic devices were there. Maybe this is why our daughter is not sleeping so well? Why our son is having difficulty focusing. And why they have both had so many colds and ear infections?

Safe in Standby?

Have you noticed that when people in an audience are asked to turn off their cell phones, most put them on mute/vibrate? This prevents annoying rings, but not the harmful exposure.

When your phone is on standby, it communicates every three to five minutes, at full power, with the nearest base station to ensure it has the best signal. If you are in a poor reception area, the phone may transmit as often as every 30 seconds as it tries to get a better signal. If you are on the move, it will transmit even more frequently.

The various modes of your mobile phone

This list sets out the various settings of a cell phone/PDA, in order of exposure from lowest to highest.

Not emitting radiation:

- battery out – Yes, there is still a tiny bit of energy coming from a mobile that is powered off. Not a concern unless you are electro-sensitive. The only way to reduce the radiation to zero is to remove the phone's battery.

Emitting radiation:

- powered off – be in this safe mode as much as you can.
- charging
- on standby – you can receive calls
- sending receiving text – less harmful if only powered on for a moment to send or receive messages.
- sending and receiving voice, online Internet, downloading movies.

Not All Networks Are EMR Equal

There are also different wireless networks, and the power of their signals varies. (In Canada, for example, one network has GSM, and all cell phones using this service emit higher levels than the same phone with the other network).

Cell Phones – Less Harmful Solutions

Obviously, the safest solution of all is not to use a cell phone, and try to avoid places where others are using them. If landline-only phone communication is not always possible for you, the following suggestions will at least reduce your exposure.

Type of phone (SAR – Specific Absorption Rate) and service

SAR is a term used to describe the rate at which tissue absorbs radiation. This is often used as an indication of the level of risk for the radiation-emitting device that is being measured.

As you can imagine, the more 'bells and whistles', the more power, the potential for more harm, we are told. Choose the most basic phone you can find. As well as taking other precautions, it may be better to choose a device with the lowest SAR possible. These SAR ratings, however, may not be reliable.

Powerwatch is a leader in this field; the organization's opinion is that SAR ratings are only minimally useful as they are based only on thermal effects. As you may remember, from Know the Evidence, the radiation from cell phones produces harmful biological effects far below thermal levels and existing government standards.

For these reasons, relying on a low SAR cell phone could give you a false sense of safety.

(LG) "The way the SAR standard is being subverted to allow more power output has been explained to me that the dimensions of the tissue measured for this test have been altered to have 1/10th the surface area. Therefore the power per unit area is reduced allowing 10 times the power for the same SAR value."

However, it still seems advisable to use a lower SAR cell phone rather than a high one (dealing with relative risk, *not* safety solutions).

As the word gets out about the evidence that mobile phones can be harmful, we can expect to see manufacturers' revised ad campaigns: "this low SAR mobile is safe ..."

Frequency of use

(RM) "**Use cell phones as little as possible. Biological effects occur after 30 seconds**. Only use your cell phone to establish contact, or for conversations lasting a few minutes, as the biological effects are directly related to the duration of exposure.

Your phone communicates at full power when it is connecting to a number, so hold it away from your body when you have finished dialing, until the person answers. This limits the power of the electromagnetic field emitted near your ear and the duration of your exposure.

For longer conversations, use a land line with a corded phone, not a cordless phone.

When you are not using your cell phone, turn it completely off (powered off, not on standby) to avoid exposure that is the result of your cell phone transmitting its location through constant signaling.

If you must have your cell on all the time, carry it in a purse, backpack or briefcase. Never carry a cell phone that is turned on in any pocket, as the radiation from the phone will couple with your body's natural electro-magnetic field and trigger whole-body adverse biological cascades.

If you must carry your cell phone on you, make sure that the keypad is positioned toward your body and the back (where the antenna is visible, or embedded) is positioned toward the outside so that the transmitted electromagnetic fields move away from you rather than through you.

Avoid having a cell phone near your body at night, under the pillow, or on a bedside table. If you need an alarm consider a much safer alternative such as a battery powered alarm clock.

Type of use

Here are some of the recommendations you need to know:

Never use a cell phone with the antenna within six inches of your body. As much as possible, use the speakerphone mode so that the antenna can be kept a safer distance away from vulnerable tissues and organs. **Never wear your cell phone in a pocket, or carry it near your heart.**

As a cardiologist, Dr. Sinatra is aware of this concern and reinforces this recommendation, "I would avoid having any wireless device near the heart; this includes a wireless microphone."

Switch sides regularly while communicating on your cell phone to spread out your exposure. It is advisable to use text messaging instead of talking when possible. Hold the phone away from you until the text is sent. Then power it off.

Is your mobile 'amping up'? The bars indicate the strength of the signal in your area; the maximum number of bars is preferable.

It's best to keep your cell phone fully charged, and best not to use it with low battery, and/or in an area of poor reception. And don't talk on the phone when the battery is being charged – that amplifies both the low frequencies (from the electrical connection) and the Information Carrying Radio Waves (the wireless communication).

Charge the phone in an area away from people and pets."

Location of use

Evidently, if you're standing somewhere and turn your cell phone on, and the base station is behind you, the signal goes through you. Yes, through you.

(CA) "When you use your cell phone in an environment where Wi-Fi or WiMAX signals are present, the body is subjected to 2 major assaults... these signals tend to couple forming new waveforms, which cause more harm."

We are also advised to refrain from using a cell phone, or other wireless device, in a car, truck, train, boat, airplane or other confined metal perimeter space.

When you're inside a metal box your mobile is continually trying to reconnect to the nearest relay antenna, and the signal is trapped inside and bounces around, concentrating the exposure and amplifying the radiation.

And it is not recommended to use your cell phone when moving at high speed, in any mode of transportation, as this automatically increases the mobile's power to a maximum as it makes repeated attempts to connect to a new antenna.

If you must have a phone on in the car, an external roof-mounted antenna will make it less harmful, though dividing your attention between a phone conversation and the road is never a good idea.

(CA) "Hands free cell phones are reportedly no safer than hand-held ones, as some studies show them being 50% more likely to cause accidents. Moreover, science shows that cell phone radiation impairs reaction time and brain function, so it is clear that the main culprit is not the 'distraction' factor – as claimed by the wireless industry – but the phone radiation which causes impairment and more accidents."

Also, please think about this: other drivers using phones may be suffering from similar impairment, so drive defensively.

Japanese physicists have also found that radiation hot spots often emerge in reflective areas, such as elevators. These researchers say passive radiation exposure is similar to second-hand smoke.

When using a mobile in a building, it is recommended to stand by a door or window and to hold it between you and the outside so less radiation goes through you as the signal blasts to connect with the nearest antenna.

1 Guideline – Limit use of your cell phone.

Have a "no frills" cell phone with speaker function – use it outside of the car, or other confined space. A wired headset is recommended if you cannot use the speakerphone.

- **Keep it turned off as much as you can.**

- Use standby mode only when expecting an urgent call.

- Turn it off – not receiving – when it's charging.

- Use a limited service plan. If you have young people using a mobile device, make sure you check the bill often. Your idea of their having a cell phone to keep them safe, and in touch with you, may not match their communication objectives. The bill will tell the true story. Remember the risk of 500 minutes a month ...

Pagers

Stan Hartman (SH) "If you have to be accessible by phone it's safer to carry a pager. Then when you are paged you can turn on your cell phone to make a quick call, if there is no landline nearby."

Remember this is "safer", reduced risk. Not "safe". Some doctors report people who have pagers have symptoms near where they carry them.

10 Safer Guidelines for Mobile Phones/PDAs

1. Use a landline whenever possible – "May I call you back on my landline?"

2. Use a low-tech, low SAR rated phone with a speaker function – avoid the hands free in-the-ear phones

3. Use on speakerphone, or with a hollow air tube headset

4. Limit calls, and keep powered off whenever possible

5. Carry away from your body – at least with the keypad or front of the cell phone facing toward the body, and the back or antenna side facing away from the body

6. Use outside of a car, train, elevator, airplane or other confined metal space

7. Use a computer with wired Internet access, with the wireless function disabled, to surf the Internet, watch movies, download music etc. not a mobile

8. Keep it fully charged (it has to 'amp' up more if the battery is weak, and/or if the reception is poor.)

9. Keep away from your sleep zone, use a battery-operated alarm clock to wake you up. Never sleep with your cell phone powered on. Charge away from humans/pets – not on 'standby' at night.

10. Internet Phones also carry risk – particularly when you're on wireless Internet access.

Wireless Earpiece Phones

The Concern:

One website marketing these all-in-one microphone and speaker wireless ear buds trumpets the advantages: **"Boosting sound and channeling it directly into the ear, reducing interference and noise."**

This may not be all they're channeling.

And, yes, these do keep your hands free for driving, which seems a good idea. However, the research indicates these devices may turn out to be the most harmful of all mobiles.

(RM) **"It is important *not* to have the speaker IN the ear, as with the microphone kind of phone.**

Magnetic fields go right through the surface so they are active in the brain at close distance."

Think about the power required to beam this pulsing digital radiation to its host or base station and then on to the cell tower – every few minutes, or so. This radiation can travel through the metal walls of your car, a concrete garage, or elevator in a high-rise building. Where is this radiation going from its snug position wedged in your ear?

In April 2008, technology writer Stewart Fist reported from Australia that lab testing of wireless earpiece phones shows **the radiation they give off exceeds current government standards.**

Wireless Earpiece Phones – Safer Solution:

Avoid these kinds of wireless hands-free phones. Use a regular low-tech cell phone with speakerphone function, or when this is not possible, use a wired, or better still an air tube, headset.

Cell Phone Headsets

The Concern:

People often ask whether using a headset is safer than holding a mobile phone right to your ear. This seems to make sense. And most experts do recommend using some kind of wired headset, if you cannot use the speakerphone.

However, **all advise against wireless headsets**. You will see the difference here again between the various levels of reducing exposure, and something being considered relatively safe.

Wireless Headsets

Many experts advise us that most headsets are not safe and that wireless listening devices worn in the ear dramatically increase whole-body radiation exposure and trigger systemic adverse effects.

(RM) "I would not recommend using a wireless headset. Even if they have relatively low power, as compared to the mobile phone, your head's exposure to the additional digital RF-signal will still be above the values recommended in 2008 by IBE (Institute for Bau-Biologie®). It also adds another signal which can increase risk."

Wired Headsets

Unfortunately, this does not seem to be the safe solution, either, as wired headsets are reducing the **near field** only. Yes, the near field is the laser-like intensity closest to your eyes and brain, but the far field is also considered harmful. Larry Gust explains:

(LG) "The wired headset does get the phone away from the head. So the big hammer is removed because you are out of the near field which is 4 times stronger than the far field. However, I have heard commentary that the radio signal follows the wire from the cell phone to the head set."

(RM) "I have concerns about using conventional wired headsets. The high frequencies emitted by the mobile phone will tend to follow the direction of the wire and "lead their way" up to the ear/head.

This is a concentrated emission and can be just as strong as holding the mobile phone directly to the ear. Also, the generation of sound in the earphone speaker produces low frequency AC magnetic fields in the range of 5 Hz to 20,000 Hz.

These fields will easily penetrate the skull since the source is the earphone, directly inside the ear. This is also why using an air tube headset is advantageous."

Headsets – Safer Solutions:

(RM) "Here is a summary of headset options – worst to best:

Worst – Cell phone held up to your ear, using a wireless headset, or a conventional wired headset

Better – The air tube headset is non-conductive thus separating you from the wires and speaker. This makes it better than a conventional wired headset. For best results, ensure the phone is at a distance and you are not in contact with the wire or the phone. Note: adapters for your particular model of phone may be challenging to find.

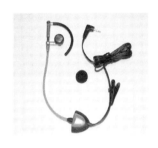

Best – Using the speaker phone function (away from the body without contacting it) is the safest option but most speaker phones have poor quality sound and do not allow privacy. This is where the air tube headset comes into play.

Safe – Standard landline phone; choose the corded phone, as often as possible. (If you are electro-sensitive, it's best to use the speaker function on the landline.)"

Every expert recommends switching from cell and cordless phones to landlines. You can get high-tech ones with call display, speakerphone function and more...

What did we learn about headsets?

(SH) "An ordinary wired headset will reduce radiation by many magnitudes. The air-tube headset is a kind of speaker that's a lot closer to your body than the speaker if you used the speakerphone function. Using the mobile phone's speakerphone is recommended."

Currently, there's no safe way to use a mobile phone, nor are there any perfect headsets. A problem with the air tube headsets, often marketed to make your cell phone safe, is that none of our experts see them as a completely safe solution. So you are reducing, not eliminating, risk.

And there's a concern that you may talk away on your mobile – cell phone and/or PDA – thinking you are now completely protected. This is the same concern with stick-on chips, discs etc. More on that later.

It seems prudent to limit calls etc. and use as suggested here. When you cannot use the speakerphone then some kind of wired headset is recommended.

I will keep this issue updated on our website. As consumers demand safer solutions, no doubt we will see more 'radiation-reducing' products on the market. Consumer beware.

The key principle that all of our experts agree upon is: if any device is communicating data through the air, wirelessly, use caution.

It's Not Just Children At Risk

When we are exposed to the radiation from mobile phones, and other wireless technology, we must consider the intermediate massive impairment of our microcirculation, and therewith the oxygen supply of our inner organs and brain, caused by the tendency of our red blood cells to stick together under the influence of radio waves (Dr. Petersohn 1998, Ritter und Wolski 2005). In addition to this Prof. Kind (Environmental Hygienic Department of Vienna University) found **a high increase of heart attacks, strokes, thromboses and embolism in people who live near transmitter masts**.

Hans Scheiner, MD

A Summary: The Radiation Rescue Communications Kit

- Corded landline phone – use for all phone calls, whenever possible
- Low-tech cell phone with speakerphone function – for quick, urgent-only calls (see Safer Use Recommendations)
- Wired (air tube?) headset
- Pager – only on at times you must be reachable
- Cable/wired Internet access

A Great Gift

Consider this kit for your family at home, and when your young people leave to set out on their own. Graduation gift?

As you know, many of this young generation are using mobile phones for all their calls and have never had a landline installed where they are living.

A young woman who works in our office is getting married, and we are giving her the first corded landline phone she has ever had since moving away from home.

Of course, it is much more convenient, and less money, not to bother installing a landline and just rely on mobile phone service.

But at what cost to our health? And, there is the consideration that a battery-operated, or not-plugged in, corded phone will keep working during power outages, and when your cell phone is not charged.

Mobile phones have their benefits, as well: great if your car breaks down. Many of our experts told me they keep a fully-charged cell phone – powered off – in the glove compartment of their car.

Video/Wireless Games

The Concern:

Many of these games use wireless technology, both on laptop computers and gaming systems. Much as you may hate the thought of trying to wrest the controller away from the young gamer, there are some concerns you need to know about.

A paper published in 2004 in the medical journal, *Human Reproduction*, reports some devices, when held close to the genitals, raise the temperature in the testicles as much as 3 degrees Celsius. Temperature changes can occur within 15 minutes. Repeated exposure can reduce fertility.

The researchers say that after a few years of daily exposure, the damage may be irreversible, and others report this exposure may be a factor in the increasing rate of testicular cancer in young men.

Computer Game Consoles

(AP) "The transformer which plugs into a power socket emits very high levels of magnetic fields, and should be unplugged when not in use. The hand controls can give off high electric fields."

Other than the considerable health concerns, there's another problem with gaming etc. It keeps our children – of all ages – inside.

Not only is this an unhealthy sedentary lifestyle, they are missing out on being outside. After the rapid-fire explosiveness of the 3-D games, strolling along the beach can seem tediously dull.

Nature, in all its wondrous simplicity, can hardly compete.

Safer Gaming Solutions:

(RM) "Hard-wired is much safer than wireless. Disable the wireless component and replace it with a cable. Encourage your children to use their game systems and laptops on a table or desk, not directly on their laps. And limit use."

(KG) "There are gaming consoles, which will still emit substantial amounts of RF radiation even when the RF antenna from the back is removed and the control devices used are wired! The popular XBOX, for example, will keep radiating as long as it is plugged in."

(AP) "It is advisable to add a grounding wire, and limit use of all video games."

Wireless Internet Access

Even the most humble coffee shop now boasts "free wireless Internet". Hard to believe, I know, but we may see a time when signs declare: "Proud to be wireless free!"

The Concern:

It used to just be cell phones. Now, with our homes, workplaces, schools and whole communities going wireless, the background level of EMR is reaching a very dangerous level.

We have already discussed in detail the mechanism of harm with this wireless radiation.

Alasdair Philips, the electronic engineer who heads Powerwatch, cautions us: "With wireless Internet access the space is filled with pulsing microwave radiation all of the time even when the computers are not in use. (Your techie may not agree.) **We believe that you should avoid using wireless Internet at all costs.**"

Safer Internet Access Solutions

Best option:

This is a situation where you really need to weigh the risks and benefits, of convenience and your health.

(AP) "It is better to get a cable or ADSL modem/router and run network cables to the rooms where you wish computers to have Internet access. This is not expensive and will avoid irradiating your living and working environment with pulsing microwaves."

(RM) "Replace wireless Internet routers with hard-wired units. Some wireless router models allow the user to shut off the wireless component and convert it to a hard-wired unit. To ensure conversion, test the area around it with a RF detector.

When using a notebook/laptop computer use a hard-wired network connection. Settings to disable the wireless connection are found in Network Settings on a PC, or AirPort on a Mac."

Chris Anderson, an EMR consultant, advises us about wireless Internet:

> As you may know, all computers nowadays are shipped with a default setting that ensures that all wireless networks and systems are turned on as soon as you start your computer.
>
> These functions need to be deliberately disabled: go to the Control Panel, click on 'wireless connections', and actually 'disable' the wLAN as well as the 'bluetooth' system.

On a personal note: during my EMR reduction frenzy, I replaced our wireless router with the hard-wired kind. Well, the phone company only had wireless ones, but the techie assured me that there would be no RF/microwave radiation if it wasn't being used wirelessly. Guess what? I had been sitting here at the computer for long hours wondering why I was getting headaches, (of course, there are always other good reasons – eye strain etc).

I decided to test the hard-wired router with a RF (radio frequency) meter. Even though I had disabled the computer's wireless settings, there were really high RF radiation levels coming from the router. Not good. Evidently, you have to go online to modify the router's settings. Now they tell me!

Another Hard-wired Option:

Networking your Ethernet and your home's power lines

If you cannot use the recommended cable hook-up – wired Internet access – you might consider this new set up where you access the Internet through electrical outlets in the wall. This is not inexpensive, but safer than having all the wireless radiation going through your house, and neighborhood.

(SH) "This is not as good as an ordinary wired modem using Ethernet cable, but way better than Wi-Fi, because the signal is going through the wiring, not through the air. Even if there is some radiation leakage from the wiring, it will be much, much weaker than the exposure from Wi-Fi."

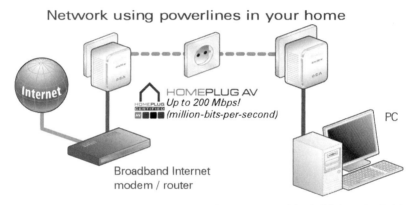

images courtesy of devolo AG (www.devolo.de)

(CA) "This option is not the best solution as some RF radiation will spew forth from the entire electrical system."

(RM) "This solution may be a bit dicey because of the dirty electricity issue, which is a bit of an unknown at this point."

Another Internet Access Option:

Have at least one computer hard-wired. If you have someone in your house who insists on using a wireless computer in another room, turn on the router for that period of time, then shut it off afterwards. **And make sure a wireless router is turned off at night**. A power bar may be helpful. You could even use one with a timer.

This safer solutions list is from an organization that I regard as the world's leading experts in this field.

How can I protect myself from wireless radiation?

The following tips help to avoid or reduce your personal radiation exposure to wireless networks:

- Always prefer wired data transmission systems to wireless networks (even though you may have to run some wire), especially in residences, schools, and preschools.

- If at all, wireless networks should be run at the lowest possible power output actually required to get the work done (which usually can be set in the software).

- Have access points only transmit for the short period of time required to transfer data, otherwise keep it always turned off, especially during the night. Simply turn the switch at the wireless device off or, better yet, unplug it.

- Preferably, keep transmitting antennas a large distance (>10m or 32.5 ft) away from areas of frequent human use.

- Transmitting antennas should not be installed in the rooms they service but rather in hallways, attics or less used rooms farther away.

- Also keep your neighbor in mind!

- Strategically place access point antennas in such a way as to minimize the radiation exposure in areas of frequent human use. Never connect sector antennas with high antenna gains (and their increased directivity) to access points and never point them towards areas of frequent human use.

- Laptop or PC: Keep your distance from the wireless card; if at any time larger amounts of data are transferred, leave the area. Always turn off the wireless card with a separate switch or disable it in the control panel (PC) or at the taskbar (Mac) when not in use.

Perhaps rooms or walls could be specifically shielded if the wireless radiation comes from outside or a neighbor; though it is imperative to have a RF radiation survey conducted prior to any shielding project.

With an external source any solid masonry wall or concrete ceiling reduces RF levels by about 50-80 %, woodframe wall constructions, however, provide hardly any reduction.

One can also cover the wireless device with shielding cloth at night or to reduce the emission level.

W. Maes, F. Mehlis: Wireless Network. A Flyer by the Building Biology Association (VB), 2009. Used with permission. (www.verband-baubiologie.de)

Wireless Systems In Schools

As the principals of my children's schools will tell you, I am a "mother warrior" on this issue (to borrow the phrase of author and autism advocate, Jenny McCarthy.)

Another concerned parent, director of a large information technology company, used this analogy:

> If someone put a glass of drinking water in front of you and said we think it's OK to drink and it meets current government standards but there is a substance in it that some credible scientists say may be carcinogenic, would you give your child a glass of that water day after day?

> The scientists and organizations in Europe that are taking positions and actions against Wi-Fi are, in my opinion, highly credible.

There is no conspiracy here; our governments and institutions have simply not kept pace with the rapid roll out of new technology. Anyone who spends a bit of time educating themselves on the issue would err on the side of caution.

Anyone who says 'I am not convinced there is a problem with wireless technology' and turns a blind eye is being imprudent. By the time they are convinced, the damage will have been done.

This issue is a dilemma for schools. On the one hand, they're told that information technology gives students an educational edge for the digital age. There's also pressure to go wireless since it's cheaper to install than hardwired systems. On the other hand, what's true of cell phone radiation is equally true of wireless Internet: children are more vulnerable to the biological damage.

The Concern:

Dr. Martin Blank is taking the lead in alerting communities:

> There is enough evidence of a plausible mechanism to link EMF exposure to increased risk of cancer, and therefore of a need to limit exposure, especially of children.

> (Radio frequency) RF in any form – cell tower radiation, Wi-Fi networks, cell phones etc. – is potentially bad, especially if one is exposed for extended periods (e.g., asleep in bed at home, seated at a desk in school).

A recent letter from Dr. Blank to a California school board is in the Resources section. Many other eminent scientists including Dr. Gerd Oberfeld, Professor Johansson and Dr. Carlo also have written about the urgent need to remove wireless Internet from schools. They cite the evidence you have read in detail in Step 1.

Dr. Magda Havas, Associate Professor, Trent University, sent an open letter to Canadian parents, teachers and school boards in May 2009:

> I am a scientist who does research on the health effects of electromagnetic radiation and I am becoming increasingly concerned that a growing number of schools are installing Wi-Fi networks and are making their school grounds available for cell phone antennas.
>
> You will be told by both the federal government (Health Canada and Industry Canada) as well as by the Wi-Fi provider that this technology is safe provided that exposures to radio frequency radiation remain below federal guidelines.
>
> This information is outdated and incorrect based on the growing number of scientific publications that are reporting adverse health and biological effects below our Safety Code 6 guidelines (see www.bioiniative.org) and the growing number of scientific and medical organizations that are asking for stricter guidelines to be enforced.

For these reasons, it is irresponsible to introduce Wi-Fi microwave radiation into a school environment where young children spend hours each day.

Dr. Sarah Starkey, a UK neuroscientist, set up an information website out of concern about the rapid spread of wireless technologies in schools. She has studied the mechanisms underlying learning and memory, circadian rhythms (like our sleep/wake cycles) epilepsy and depression.

According to research on www.wifiinschools.org.uk many of the health risks may only become apparent after years of exposure. Some students and staff could be susceptible to more immediate effects, such as symptoms of electro-hypersensitivity. There are studies that suggest that people with epilepsy have an increased chance of seizures when they're exposed to cell phone signals.

Wireless Networks, Immune Systems and Viruses

Do you remember that Dr. Scheiner, and other medical experts, have reported wireless radiation suppresses the immune system? With the current spread of the H1N1 virus, and its threat to schools, wouldn't you think that school principals would be more than keen to go back to hard-wired Internet connections and reduce all other exposure levels, as much as possible?

These precautionary steps also make sense in other public spaces, including confined ones like trains, buses and airplanes. Also, you'll read later on what one of our scientists, an immunologist, has to say about wireless radiation and the immune system, and what he recommends if you have to have a flu vaccine (to keep you from skipping ahead, let me tell you here: he suggests the nasal mist, that is mercury-free, instead of the injectable vaccine.)

Educational edge – or irony?

A school principal has made a pitch to the school Board to purchase a wireless Internet network to enhance his students' academic performance. He wants to give them every high-tech advantage. He has read credible research of the many educational benefits of computer technology: better test scores, better learning, better focusing. With the wireless Internet network, and a wireless laptop for each child, he is confident these children will have the leading edge.

One of the moms is alarmed that her children seem to be even more hyperactive – alternating with fatigue, and occasionally not sleeping well. Sharing her concern with other moms, she discovers their children are also having trouble concentrating; some are experiencing symptoms like headaches, dizziness, nausea, diminishing eyesight, tingling in the hands and impaired immune systems. Two are having difficulty recovering from mononucleosis and Lyme disease.

One parent remarks there were never this many allergies when she was in school. Some test scores are dropping. And there seem to be more 'problem children' who simply can't sit still and focus on their work. These mothers turn their formidable energies toward researching the possible causes – and solutions.

Decrease In Memory, Attention and Ability To Learn?

Electro-magnetic radiation can alter the electrical activity in the brain – with unknown effects on the still-developing brains of children and teens.

Some studies also show a decrease in cognitive abilities such as memory, attention and decision-making.

Does it makes sense to radiate an educational environment when this exposure has been shown to damage our children, including their ability to sit still, and to learn?

(AP) "School libraries and IT facilities usually have wireless local area networked (wLAN) computers. These can give a high background level of microwaves to which teachers and pupils will be exposed for most of the day. We are against every pupil having a wireless enabled computer to do his or her school work.

This is because 20 or more computers, with wireless access, fill the classroom with several volts per metre of pulsing microwave radiation electrosmog. Electrically sensitive children will react very badly to this exposure, which is associated with memory and concentration problems and increased aggressive behaviour."

(AP) "**Laptops wired by Ethernet using standard cables are fine.** The pupils can use their laptops on their own for most work and plug in a network cable when they need to access files on the school server computer or on the Internet. Infrared data connections are also believed to be safe."

(AP) "Interactive whiteboards, used to display computer and video images in the classroom also use microwave technology. The whiteboard display (using back or front projection) should be connected by wire to the teacher's controller PC and will not then be a problem.

However, some are now being sold with built-in wireless networks to connect to the teacher's PC and to wireless connected interactive pen displays that can be moved around the classroom and operated by pupils.

These wireless options will increase the ambient level of electrosmog in the classroom."

Wireless Internet In Schools – Safer Solutions:

The obvious solution is to remove the wireless network and go back to the hard-wired system. Dr. Starkey advises parents to take a stand on wireless in their children's classrooms:

> Schools will only use wireless computer networks if the parents want this for their children. Obviously it is not desirable for a few children to be isolated, or to miss out in their education.
>
> But if parents en masse decide together that their children will only use wired technologies then schools will need to revise their communications technology provision.

Reducing your child's EMR exposure

- Minimise the time your children play computer games.
- Monitor the type of games they play.
- Keep children at least a metre (3.25 feet) away from working electrical appliances in the kitchen.
- Make sure your child sits at least a metre (3.25 feet) away from the front of a TV screen while watching.
- Position the television set so children cannot spend time to the side or back of it.
- Make sure transformers are kept away from chairs, beds and anywhere a child sits and plays regularly.
- Consider your use of battery-operated equipment. Use rechargeable batteries where possible and dispose of them as toxic waste.
- Position your child's bed away from cupboards containing electric equipment, meters, heaters and the like.
- Make sure all music/entertainment players are plugged in with a grounded cord.
- Make sure your child sits well away from the transmitting unit of a cordless headphone system.
- Make sure desk and bedside table lamps are grounded.
- Check your garden for EMR sources and build play areas where field levels are low.
- Make sure your child sits in seats that are not exposed to high EMR in public transport.
- Check your child's nursery and school for sources of EMR.
- Monitor your child's health, behaviour and achievement levels and check school EMR exposure if these deteriorate.

Check whether your child's school has a wireless system or interactive whiteboards. Measure the EMR in the classroom to see if it is at an acceptable level. (Source: Powerwatch)

Fluorescent Lights

Most people working in offices with overhead fluorescents, whether they are buzzing or not, report eyestrain, fatigue and headaches. The recent promotion of compact fluorescent lights for their energy efficiency causes the scientific community concern. There are two problems with the compact fluorescent lights – one EMR assessment technician calls the CFL "chronic fatigue lights" for their negative impact on our energy. The second problem is they feed what is called dirty electricity back into the electrical system.

Dr. Magda Havas (MH): "Don't use CFL bulbs. These energy efficient compact fluorescent lights generate radio frequency radiation, ultraviolet radiation, and dirty electricity. Dirty electricity has been shown to adversely affect human health. A recent study of cancer clusters in a school in California associated the increased risk of cancer among teachers to dirty electricity. **CFLs contain mercury – a known neurotoxin** – which can be dangerous once it is released into the environment through breakage or in landfills. With thousands of bulbs sold around the country and inadequate disposal facilities we are setting ourselves up for a mercury time bomb. Compact fluorescent lights are making some people ill, including those who suffer from migraines, epilepsy, skin problems and electrical sensitivity."

Many environmental groups, and governments, are advocating a ban on incandescent lights and mandating the use of CFLs. When you read the science it seems evident we need to find safer ways to cut our energy costs and consumption.

(AP) "Many schools (and workplaces) use fluorescent lighting. The flicker can trigger electrical hypersensitivity and increase antisocial behaviour. If your child develops headaches and/or eye strain, check the school classroom lighting."

Fluorescent Lights Safer Solutions:

(AP) "Ensure lights are full-spectrum – not fluorescent – as it has been found that people are much healthier and thus able to learn better (and work better) in lighting that approximates daylight as closely as possible."

(KG) "If using fluorescent lights – standard or full-spectrum – be sure to have them electrically shielded with a proper grounding connection."

(RM) "Incandescent light bulbs are a better alternative. LED (light emitting diode) technology looks promising as well."

(MH) "Instead of promoting compact fluorescent light bulbs, governments around the world should be insisting that manufacturers produce light bulbs that are electro-magnetically clean and contain no toxic chemicals. LEDs are a safer option."

Unlike compact fluorescent bulbs, LEDs contain no mercury and they work well in cold weather. They may provide a more pleasing light than fluorescents and appear to be the greenest and potentially healthiest option.

(KG) "Watch out for poor quality electronic drivers in LEDs that increase electro-magnetic field emissions and may cause flickers. Models with a proper grounding connection are preferable."

The high cost of LED light production has meant that their use has been limited to commercial and outdoor lighting, but that is changing. Soon they will be more available, and more economical, for use in our homes, offices and schools.

Halogen Lights

The Concern:

(AP) "Low voltage halogen lights are also not recommended as they give off high electric fields, and many have their own built-in transformer. This is usually poorly constructed and gives off very high levels of power-frequency magnetic fields close to them."

Halogen Lights Safer Solutions:

(KG) "If using low-voltage halogen lights, the transformers should be located away from high-occupancy areas and the wiring shielded. Since most transformers in low-voltage desk lamps emit magnetic fields even when turned off, avoid these. Line-voltage (120V) halogen lamps are better, especially for bedside and desk lamps. See the lighting recommendations in my paper on low-emission office environments: www.buildingbiology.ca/healthyoffice.php"

Night lights

The Concern:

(AP) "There is plenty of strong scientific evidence that shows that even quite low levels of light stop the production of melatonin, a hormonal chemical synthesised in the pineal gland. It, and its associated chemicals, 'mop up' dangerous free-radical damage. There is good evidence that low levels of melatonin increase the likelihood of cancers and other serious health problems developing. It is often your anxiety about seeing your child that starts the 'light at night' habit."

Night Lights – Safer Solutions:

The only type of night light that you should use are the very low power (usually plug mounted) orange or red ones that just glow gently as these colors hardly affect the production of melatonin. White and blue-white lights are likely to stop the pineal gland's production of melatonin for the whole of the night.

(KG) "I am very concerned about light exposure at night, no matter what its color. Darkness is best."

Dimmer Lights

The Concern:

(RM) "Dimmer switches produce elevated magnetic fields and power line transients, which are high frequency noise on the power lines."

Dimmer Lights – Safer Solution:

(RM) "Remove dimmer switches."

(MB) "Get rid of instant-on devices."

(SH) "Do not have dimmer lights or touch-on (clap-on) lamps near where you sit or sleep."

Even if you choose to be wary of cell phones, cordless phones and wireless Internet you are still be exposed to EMR in all kinds of ways that have never occurred to you, and ones that you may not be able to control.

Consider the following.

Medical Technologies

One of Dr. Becker's concerns was that our conventional medical system relies on powerful testing and treatment technologies that may administer high doses of EMR. Dr. William Rea, also a surgeon and a highly respected pioneer in this field, reports that he became electro-sensitive from being immersed many hours a day in the radiation around him.

At what point do the risks begin to outweigh the benefits?

Pre-natal ultrasound

This snapshot of your unborn child can offer a magical moment for expectant parents, and also crucial medical information. I just want you to know that some experts have advised caution.

How it works: ultrasound is a sound wave transmitted in water or biological tissue. A low energy ultrasound is used in medicine as a diagnostic imaging technique, especially in obstetrics.

The Benefits: finding out whether a fetus is growing normally.

The Concern:

Some expectant parents are subjecting their babies to multiple ultrasounds for reasons that aren't medically necessary. The Federal Drug Administration (FDA) in the US, and the Society of Obstetricians and Gynecologists of Canada are among the medical experts who caution:

Although there have been no fetal abnormalities linked directly to diagnostic ultrasound, it involves targeted energy exposure to the fetus, and there remains a theoretical risk for subtle effects on fetal development. This is suggested by some biological effects of ultrasound observed at, or near, diagnostic intensities in both human studies and animal models.

Source: Non-medical Use of Ultrasound – SOGC Policy Statement April 2007

There is also concern that some technicians may deliver a stronger exposure than required, or may try to reposition the fetus for a better viewing angle.

(KG) "The sound level of an ultrasound scanner is 100 dB; in the human audible range, this would be comparable to subjecting your fetus to the sound level of a jackhammer at 1 m distance."

Ultrasound Safer Solution:

The experts' recommendation: have ultrasound only when medically necessary.

Dental/Medical X-rays

Electro-magnetic waves at this very high frequency – called ionizing radiation – may damage or kill cells through the production of free radicals.

What are free radicals?

You'll need a small dose of science here if you want to know what this means. The high-energy waves from X-rays interact with

DNA in your body tissue and dislodge electrons, splitting the atom into negatively charged electrons and a positively charged ion.

This ion, or free radical, behaves wildly seeking to replace its missing electron. (I once heard this described as 'recently single, seeking a mate'.) The unleashed reaction of free radicals in the body can lead to the disruption of intra-cellular metabolism and interference with DNA repair.

As you may know, many forms of cancer are thought to be the result of free radicals reacting with DNA, resulting in unwanted cell mutations. (Antioxidant nutritional supplements are considered by many physicians to reduce the production of these free radicals.)

Dental X-rays

Again this procedure has its benefits – and risks. I was often told that there was no concern about these levels of radiation. ("Just like a day in the sunshine".) Now I'm not so sure this is true.

Our new family dentist does not have the digital X-ray machine, but I found one office that does, so if any of us need a dental X-ray we go there to have this lower-exposure kind.

How dental X-rays work: a beam of radiation is focused to go through your teeth and onto a small piece of photographic film. Dense structures such as teeth and bones absorb more X-rays than the soft tissues. Cavities and infections will show up as lighter areas because more radiation has passed through them before striking the film.

The Benefit: finding decay and diagnosing periodontal disease in impossible-to-see areas, such as under fillings or between teeth.

The Concern:

While the amount of radiation in any one X-ray is reportedly 'small', exposure is cumulative – it builds up with each one. In the head and neck, areas exposed to dental radiation, X-rays increase the risk of damage to, or cancer of, the lens of the eye, thyroid, salivary glands, bone marrow and skin. Scientists in the UK have found that about 0.6% of the cumulative cancer risk in people under age 75 – about 700 cases per year – could be attributed to diagnostic X-rays. In Japan, which has the highest frequency of diagnostic X-ray exposure in the world, the risk was more than 3%.

Source: The Lancet, Vol. 363, Issue 9406

Dental X-rays – Safer Solutions:

Question your dentist – is the X-ray essential? Why? I switched dentists when one insisted our children needed whole mouth X-rays "on file". Find a practitioner who uses newer digital X-rays, which evidently beam out 90% less radiation.

Medical X-rays – Safer Solutions:

It's a similar situation with medical X-rays, used, among other reasons, to check for broken bones, or in the case of chest X-rays, to diagnose pneumonia. I don't want to scare you away from essential medical tests, but to encourage you to discuss with your practitioner whether there's another way to get the information needed for diagnosis.

Again, is this X-ray essential? Do people with recurring sinus infections really need to have their heads X-rayed? What is the effect on their eyes? Brains?

CT and MRI scans

I also hesitate to sound an alarm about these tests, as it seems unlikely anyone would have them unless it was medically crucial – when the benefits would outweigh the risks. Or so I used to think! Private clinics are now marketing full-body CT and MRI scans to the general public, even those who have no other signs of disease.

Health authorities agree there is no scientific evidence that using these scans on people without symptoms is useful in catching diseases early. If the tests produce suspicious findings, this could lead to having more unnecessary procedures, exploratory surgery, medications, etc. If the scan shows nothing, you might then ignore signs of an actual illness that can't be detected by scans. And that's before you even consider the potential side effects of the procedures themselves.

The Concern:

In December 2008, American researchers, developing tools to help doctors weigh the risks and benefits of CT (computed tomography) scans, reported that as many as 7 percent of patients from a large U.S. hospital system received enough radiation exposure from scans over the course of their lifetime to slightly increase their risk of cancer.

The researchers raised the concern that repeated scans of people who had cancer could increase the patients' chances of developing another tumor.

And in February 2009 a paper in the Journal of the *American Medical Association* reported that one CT angiogram, widely used by cardiologists to find potentially fatal blockages in heart arteries, could deliver a radiation dose equivalent to 600 chest X-rays.

Researchers caution that whole body CT screening exposes you to radiation levels that up to 1000 times as high as those from an X-ray, increasing your risk of developing radiation-induced cancer.

MRI (magnetic resonance imaging) is a different technology, as you may know. Its powerful magnetic field excites the molecules of the body's tissues, causing the nuclei in these molecules to emit a radio frequency. The MRI scanner transmits these radio signals to a computer, which translates them into a 3-D image of the area being examined.

Unlike CT scans, or other kinds of X-rays, this process does not involve the highest power – ionizing – radiation. Some companies are marketing MRIs as having 'no radiation'; this is not completely true. And, some people claim they became electro-sensitive only after having had an MRI.

Unnecessary scans could increase the risk of bringing on the very diseases you're worried about and trying to detect early. So the recommendation is to have these tests only when medically necessary. Even if your doctor is ordering them, you have the right to question. And to a second opinion.

Mark Baerlocher, MD, a resident in Diagnostic Radiology at the University of Toronto, offers his concern about, "the lack of awareness, and the general misinformation regarding the radiation risks associated with Medical Imaging and related exams, and procedures, among both healthcare professionals and patients." This concern led him to develop the Radiation Passport, a computer program to help patients, physicians, and dentists track lifetime exposures and better understand the risks. One can look up the estimated effective radiation dose associated with a given Medical Imaging exam or procedure, and the associated estimated risk of developing cancer due to that radiation.

Medical Scans – Safer Solutions:

Cardiologists say CT angiograms are indispensable in some cases. I was relieved to read that techniques are being developed to reduce the amount of radiation given in these valuable tests. Some physicians who practice complementary medicine recommend a protocol to reduce the impact of X-rays and scans. See *Prescription for Nutritional Healing,* Avery, New York, by James Balch, MD and Phyllis A. Balch, CNC.

Other Exposures:

Radio Frequency ID tags/wristbands

Wireless chips are already everywhere: in tags on clothing in stores, in security access cards, some passports, tollbooths, library books and bus passes. Recently, I saw little units being handed out at a pharmacy prescription counter so the staff could beep people when their prescriptions were ready. I walked past the beeper station quite quickly.

Many of these devices only emit signals when they're run through a scanner. But the ones that are causing more concern are those that send out signals constantly.

How they work:

The chips "communicate" with each other by generating radio frequencies. In hospitals they're used to track medical supplies, equipment, and even people. One hospital in Israel, for instance, was using RFID wristbands on babies in the neonatal ICU, to make sure each preemie was fed from a container of his or her own mother's milk. (Someone please tell them.)

The Benefit:

Better inventory control, faster service. In hospitals, RFID bracelets can help keep track of people with dementia, for instance, who might wander off.

The Concern:

According to a study published in the June 2008 Journal of the American Medical Association, tests show these devices may interfere with medical devices such as pacemakers, ventilators, IV pumps and anesthesia machines. The RFID tags used in hospitals are the "passive" kind – no internal power supply and need a "reader" device to detect their radio waves. The "active" tags, as in retail stores, are constantly broadcasting their signal. What happens if someone with an implanted pacemaker tries on a piece of tagged clothing?

Even though evidence exists that tumors grow at the site of chip implantation, some pets are still tagged this way.

ID Tags – Safer Solutions:

Hospitals should check all new technology, especially wireless chips, to make sure they don't interfere with medical equipment, and to know about potential adverse biological effects – in their patients, and staff. It is also possible to shield the RFID chips so that they're less likely to cause problems.

If you have an implanted defibrillator or pacemaker you might want to pay attention to where these devices are in your environment.

Airport and Other Security Full-body Scans

How they work:

These machines use X-rays to electronically "strip search" passengers. These scans produce an image of people's bodies, and any weapons, explosives or illegal drugs they might by carrying.

The Benefit:

Avoiding actual strip searches, speeding up airport security processes, and theoretically keeping us all safer.

The Concern:

Exposing travellers to radiation. Some radiologists, including one interviewed by BBC News, are concerned about the potential effects on women in their first trimester of pregnancy. Radiation exposure can damage reproductive DNA – risking genetic abnormalities in the fetus.

(KG) "During pregnancy or with babies, it is best not to travel by airplane, certainly not on an intercontinental flight. The ionizing radiation exposure from the cosmic radiation is rather high, especially the higher the altitude of the flight."

Security Scans – Safer Solutions:

If you're a frequent flyer and you don't want to be irradiated, you may have to put up with being patted down by hand. If you are electro-sensitive or pregnant, or even think you might be, inform security staff. After you identify your exposures, your next step to reduce them as much as you can. It's not always easy, I know.

Smart Meter

A "Smart Meter" is a remote way the electric, gas or water company reads a meter. There are several ways these meters work. Some use radio waves that can cause some people to feel ill. Find out what method is used for your meter. If it is a radio wave method, ask that the Smart Meter be replaced with a regular mechanical meter. You may have to pay a bit more, but your exposure is less.

Wired Devices

Now let's look at the devices from the lower frequencies – the first part of the Spectrum. (As you may remember, we examined the third category – X-rays, etc. – in our last section with the Risk Survey information.)

Computers

Computer screens are an EMR issue.

As you may know, the older standard monitors, with cathode ray tubes (CRT), emit high levels all around them. These are the curved back chunky kinds, sadly often found in school computer labs.

All CRT monitors give off higher fields at the back and sides. Always sit more than a metre (39 inches) from the rear. I have been told by EMR technicians that if you have a bank of computers, it is better to put them back to back to minimize this exposure.

Magnetic fields travel through walls, so watch the use of the next room (for instance, don't let it be your child's sleeping room).

Smaller monitors are not necessarily better than large ones, because the field's strength depends more on the internal design than on the screen size.

The flat LCD screen is a better choice. I chose the low-tech LCD model that had small speakers and no other frills. It also has a good-sized screen so that I could place it far back from where I sit - more than an arm's length away.

(SH) "Infrared computer gear is okay, including some keyboard and mouse connections."

With the keyboard, again you want the wired, not wireless. The same applies to the mouse. Also, I have heard that the optical mouse, because of its design, has a higher electro-magnetic field than the laser variety. I am using a corded laser mouse. If I find ones with proven lower EMR levels, I will post this information on our website.

Laptop computers

The Concern:

(AP) "These handy computers can emit significant levels of electric fields from the back illumination and scanning processes. When run from the plug-in adapters, they can give off very high electric fields next to the keyboard and display, of several hundred volts/metre."

Not too long ago, a woman contacted me when she heard about this book, asking if the numbness in her legs might be caused by sitting for long periods with her wireless laptop on her lap. Evidently she did a trial of sitting in the same position for the same amount of time without the laptop, with no numbness in her legs.

When I gave her the short story about effects of wireless technology, she responded abruptly, "Oh, I couldn't do without that. I'm a writer."

Laptops – Safer Solutions:

(KG) "Laptops also give off very high magnetic fields at close range where you have your hands at the keypad, therefore it is better to use a laptop with an external keyboard whenever possible."

(AP) "Charge laptop computers away from where you sit, and then run them off their internal recharged batteries, or with a grounded power supply."

Remember what I said about proximity?

This might sound contradictory, but all of the experts say don't use a laptop on your lap. Whenever possible, place it on a desk or table or with some other surface between the computer and your body.

I have also learned that electrical devices react with each other, even through walls, so you don't want to sit between your computer and in the direct line of fire with another strong electrical device, like a television, for example.

There are many good reasons to limit computer time; as you know gazing absentmindedly, or even mindfully, at a computer screen can be a factor in a myriad of eye symptoms and conditions. Again, there is a real concern with the amount of time children spend staring into computer/video screens.

Encourage your children to get outside, whenever you can. Maybe even wander outside with them. Doesn't matter if it's not a bright sunny day. As Richard Louv wisely remarks in *Last Child in the Woods,* "There's no bad weather, just wearing the wrong clothes."

With so much time sitting at a computer, many of us suffer from repetitive strain injuries, such as carpal tunnel syndrome, headaches and muscle pains in the low back, shoulders and neck. All in all, it's not the best activity for our wellbeing.

On my computer I keep a sticky note reminding me to:

Stretch. Breathe. Blink. Take a Break. Get outside.

In your office at home or at work, there are many electrical devices you may want to move farther away from you, downgrade to a lower tech version, or remove altogether.

Other office equipment

How many times a day do you see people clustered around a huge, humming photocopier waiting for a long document?

The concern:

(AP) "Photocopiers give off quite high levels of EMFs from the motor. It would be advisable not to stand immediately next to the motor while it is working. Photocopiers also emit significant levels of ozone, and should always be in a well-ventilated room, or in a corridor. Volatile organic compounds (VOCs) are released, which increase with warmth, and double-sided copying. VOCs are given off even when the photocopier is not working, but is still switched on."

(AP) "Printers, scanners and fax machines can all give off significant levels of EMF near them, and some laser printers also release ozone.

Office Equipment – Safer Solutions:

It's obviously best to place printers, fax machines, copiers etc. as far away as possible from where you sit, and don't stand next to them while they're in operation. This is the basic recommendation with all electrical equipment.

If you have a wireless printer, consider replacing it with a wired one. Ink-jet printers are more ecologically friendly than laser printers and do not give off ozone.

Entertainment Devices

The Concern and Safer Solutions:

(AP) "**Televisions** give off high magnetic and electric fields up to 1.5 metres away (about 4 feet). Remember that magnetic fields travel through walls. EMR from **digital TVs** is not much different in level from older analog models, though some people with EHS say that they are more affected by them. Children should not sit close to any size television. Watch TV, including plasma screens, from a minimum of 2 metres (7 feet) away.

Plasma screens give off high levels of electric fields. Use the main switch on the set to switch off the television. Some types of remote control leave the TV on standby and it continues to consume up to a quarter of the energy it uses when switched on."

(KG) "LCD TV screens usually have much lower EMF emissions than plasma TV screens."

These days, all TV and video remote controls work using very low-power infrared light, and pose no problems.

(RM) "Some high tech remote controls do use RF but are only transmitting when buttons are pushed."

Analog TV reception signals are relatively low power and are believed to have few biological effects.

(KG) "It is true that digital TV radiation seems to be more biologically active than analog TV. But if you happen to live in the vicinity of an analog TV tower, levels are so high that this will also cause major adverse health impacts."

(RM) "TV reception signals are switching to digital and these will pose a health threat."

Digital TV receivers can give off high electric fields if the TV system, or satellite decoder, is not connected to the building's electrical grounding wire. Most TVs, video recorders and satellite systems are not grounded when you buy them, as they only have a two-wire power cord, rather than a grounded three-prong plug. Walls will give some protection from the electric fields; windows are less effective at screening them.

DVD and video players and recorders have transformers, in their power supplies, which can generate strong magnetic fields.

TV, video and Hi-Fi microwave transmitter boxes, used to transmit pictures and sound from your main entertainment system to all other rooms in the house with a suitable receiver box, should be avoided. They are another source of microwave exposure in your home environment.

Personal portable music players

(RM) "Personal music players can affect the electrically sensitive even if not Wi-Fi. Recommendations are to keep electronics off of the body."

Note: The new 'improved' wireless-enabled music players are not recommended for the obvious reason that the listener is now exposed to wireless radiation. Listening to music from a laptop, which is wirelessly connected, is also not recommended.

Electric Heating Pads, Electric Blankets, Electric Mattress Pads, etc.

Electric blankets and heating pads seem comforting, and health-enhancing. Or at least, we think they are. Remember my surrounding my dear mother with this electric warmth to keep her sluggish circulation moving?

The Concern:

(RM) "These heating devices have wires threaded through them, and electricity is used to create heat. I think it is easy to see the issues when they are plugged in and turned on. While operational, extreme levels of AC Electric and possibly AC Magnetic Fields can exist.

If they are turned off and still plugged in there still may be a high electric field because of the mere presence of electricity on its power cord. And the wires that run throughout the items can act as an antenna for radio waves and also re-radiate other electric fields from other sources."

Dr. Andrew Weil, (AW), a respected physician and health educator advises on his website: www.drweil.com:

(AW) "Household appliances such as heating pads, electric blankets and mattress covers, plug-in hair dryers, computers, and coffee-makers all generate EMFs that surround electrical equipment, power cords, and power lines.

Some research has suggested that exposure to low-level magnetic fields emitted by such appliances as hairdryers, electric blankets and electric razors can damage DNA in brain cells.

The data come from a study in rats at the University of Washington (conducted by Henry Lai, PhD) which found DNA damage in animals exposed to a 60 hertz field for 24 hours; more damage was found after 48 hours."

Electric Blankets, etc. – Safer Solutions:

On his website, Dr. Weil also suggests: "Because we can't say for sure that electro-magnetic fields pose no risks, I would advise you to use a non-electric heating pad, such as one that you warm up in hot water."

(RM) "Find non-electrical methods for heating your bed, like an organic wool/cotton comforter. And it is recommended to keep wires and metal away from the sleeping area. Also use a non-metallic mattress and foundation like natural latex with an organic cotton/wool topper."

Your Sleep Zone – Another Reminder

As we have mentioned, all experts advise making our sleeping areas electronic-free zones, as much as we can. This is especially important if anyone is vulnerable due to age, illness, electro-sensitivity etc., or who is experiencing any of the symptoms or conditions we have listed as being related to EMR exposure.

If anyone is having trouble sleeping, lowering the levels, as much as possible, is strongly recommended. Most electro-sensitive people have sleep disorders. Meters can help determine the best place in the house to sleep and to monitor the effectiveness of interventions.

While having a low EMR sleep zone is very important for us all, it is *crucial* for babies and young children.

Baby's Room – Must be a low-EMR environment

(AP) "Arrange your baby's sleeping area so that he or she is at least a metre (3.25 feet) from the nearest electric appliance and about that far away from the wall. Don't use a dimmer switch to provide a light in the baby's sleeping room, unless it is well away from the baby's head. Ensure the bulb it is lighting is red or orange, not white or bluish."

You know, of course, that it is prudent to avoid electronics in a baby's, or young child's room, as much as possible. These precautions include avoiding plugged-in clocks [use battery ones], fluorescent lights [use incandescents] and avoiding cordless phones.

It is also important, if you have young children in the house, to make sure your Internet access is wired, not wireless, as a wireless router radiates for hundreds of feet (metres).

(RM) "Be aware of what is in the room below or beside the baby's sleeping area. I find many houses have fluorescent light fixtures mounted underneath the bedrooms, which may be on when the children are sleeping. Make sure no one, especially a child, is sleeping close to appliances in adjoining rooms."

(CA) "I would also recommend not having televisions or other electronics on the other side of the wall from where a crib is located."

Baby Monitors

The Concern:

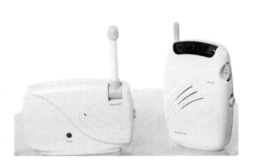

Now that you know a lot about the health hazards of wireless radiation and digital DECT cordless phones, which cause our experts so much concern, imagine putting one of these high-powered devices near a newborn child.

If you know parents who have digital/wireless monitors in their child's room, please let them know. Let's alert the companies that make these, and the retail and online stores that sell them.

When the manufacturers hear that we won't buy digital/wireless monitors they will bring back the much safer analog/wired ones.

Larry Gust comments:

(LG) "The baby monitor signal used to be run through the house electrical wiring to where mom or dad were listening.

Then progress was made and the monitor went wireless using

analog radio waves. This was okay as so long as the monitor was not placed within six feet of baby.

Then more progress, and suddenly we are using digital radio signals. A component of this signal is unhealthy for the cells in the baby's body (your body too).

Now virtually all baby monitors are wireless AND digital – emitting digital microwave radiation throughout the baby's room."

(KG) "Digital baby monitors measured one foot away (30cm) emit RF levels up to 20,000 (microwatts per sq. metre)."

Baby Monitors – Safer Solutions:

(AP) "Avoid the 2.4 GHz (gigaHertz) and 5.8 GHz digital baby monitors. Use the old fashioned plug-in type or the older analog type monitor. As long as the analog monitor unit is six feet away from the baby, it does not appear to present a risk.

I encourage people to try and find older ones – at garage sales, thrift stores, eBay, etc.

Look for the antenna to be about 6 inches or more long. If the antenna is not visible, or is one inch or less, the unit is the 2.4 GHz or 5.8 GHz digital type and is to be avoided."

Baby Sensor Pads/Bedding

(AP) "Sensor pads are put under the mattress with the intention of monitoring the baby's breathing. One well-known UK make was found to produce significant levels of pulsing microwaves around the baby's body."

(LG) "Some parents may use sensor pads as an extra protection against crib death. Dr Jim Sprott, OBE, a New Zealand scientist and chemist, states with certainty that SIDS is caused by toxic gases. These gases can be generated from a baby's mattress when a fungus, that commonly grows in bedding, interacts with fire retardant chemicals that are added to mattresses (Richardson 1994). Depending on conditions, these heavier-than-air gases, generated by the fungus acting on the retardants, are concentrated in a thin layer on the baby's mattress, or dissipated into the surrounding atmosphere. If a baby breathes or absorbs a lethal dose of the gases, the central nervous system shuts down, stopping breathing and heart function. These gases can fatally poison a baby, without waking the sleeping baby and without any struggle by the baby. A normal autopsy would not reveal any sign that the baby was poisoned (Sprott 1996 & 2000)."

Baby Sensor Pads/Bedding – Safer Solutions:

(LG) "The fundamental solution is to eliminate all sources of phosphorus, arsenic and antimony from mattresses. But this is not happening now. However, putting the mattresses in a gas-impermeable cover made from high-grade polyethylene can prevent exposure to these gases plus insuring that bedding used on top of a covered mattress does not contain any phosphorus, arsenic or antimony.

Dr. Sprott specifies a fleecy, pure cotton (flannelette) under blanket, with only cotton, or poly-cotton, sheets and woolen, or cotton, blankets over the baby. No other bedding should be used in the baby's crib. In particular, do not use any synthetic sheets or blankets, duvet, sleeping bag, or sheepskin (Sprott 1996)."

Crib toys?

Battery-operated stuffies/toys look so cuddly – some even emit a range of soothing sounds – but are they safe?

(LG) "The level of radiation is very small. Yet, newborns are in a critical phase of life; they are not hardy. I would adopt a conservative attitude and not give these to an infant."

(AP) "I would have these toys outside of the bed/cot. And I see no good reason that sound generators need to be built inside a toy that the child has. Be aware, that increasing numbers of these toys have a microprocessor inside that will generate some EMF very close by."

Hair Dryers, Electric Razors

The Concern:

If you are a barber or professional hair stylist, this is a significant concern. Remember the factors of intensity, proximity and duration?

Dr. Henry Lai cautioned:

> Data suggest that the effects of EMF exposure are cumulative and may build up in humans over time as a result of repeated brief use of common plug-in appliances. I suggest limiting exposure to as little time as possible, particularly with devices used close to the body.

(AP) "The motor of a hairdryer gives off very high fields near the handle, dropping only a little at normal drying distances (6 to 18 inches). The fields are slightly higher on high heat, than low heat. Any metal hairpins will increase exposure. It's recommended not to use hair dryers etc. after 7 pm. High magnetic fields near the head in the evening could interfere with the production of melatonin – needed for sleep – by the pineal gland for the rest of that night.

The heating units for electric hair curlers give off significant levels. Sit away from the heating unit while it is still working. When you have removed the curlers, switch off the heating unit.

Hair curling tongs and flattening irons give off quite high fields. It is best to use them with care.

Sitting under a hooded hairdryer regularly is not recommended as they emit a high field. Metal pins will increase EMR exposure.

Electric razors also give off high fields. Again, this is more of a concern with professionals who are handling them all day."

Having read this far, no doubt you are now leaping ahead to your own safer solutions considering proximity, duration etc. There are also lower-EMF hairdryers advertised. What meter would you use to test the field around them? Yes, the Gauss meter that measures the non-communicating type of radiation.

Hair dryers, Electric Razors – Safer Solutions:

(AW) "The strength of EMF falls off exponentially as distance from the source increases. So hold the hairdryers farther away."

Non-electric razors are an option if you are concerned about this exposure.

Electric Alarm Clocks/Clock radios

Imagine my consternation when I saw a fully-loaded bedside unit: a sleek silver wireless router, alarm clock and CD combo unit for sale, in a pharmacy, no less!

The Concern:

If a plugged in clock, or any other electronic gadget, is lying within a few feet of your head all night, you're receiving that exposure when your body most needs a low-EMR environment. This is not a minor concern and may have significant impact on our health over time.

Clocks – Safer Solutions:

(SH) "Plugged-in clocks, radios etc. should be kept at least 6 feet away from where you sleep."

The best option is a battery-operated alarm clock. The next best is a small electric alarm clock placed at least six feet from you; those of us older than 40 may have to get one with larger numbers!

Other Electric Equipment

Hot Tubs/Whirlpool Baths – Yes, these tubs of hot swirling water are relaxing, but we should know the powerful motors produce electro-magnetic fields.

(AP) "Most have pumps and motors built in to the base resulting in high EMR exposure as you take a bath. It's recommended that the pumps and motors are at least half a metre (roughly 2 feet) away from the bath.

Sunbeds emit high electric and magnetic fields, as well as possibly dangerous levels of ultraviolet radiation. And many can give off five times as much UVA as would be expected from bright sunlight at the equator.

These will increase your risk of developing skin cancer, especially if you are fair-skinned.

Sunlamps – the EMR is not as high as sunbeds, but like sunbeds they give off ultraviolet light – a form of radiation that we know causes skin cancers.

Exercise machines – Motors (such as those used to power treadmills) give off high magnetic fields. You are usually not that close to the motor so they are generally not a problem. The exercise is probably more beneficial than the small level of potential risk.

Vibrating muscle toners emit both electric and magnetic fields. Probably all right if you are not electrically sensitive or pregnant."

(KG) "Depending on the device, muscle-toning machines can emit substantial amounts of EMF. A rebounder is definitely a safer option."

(AP) "**TENS** (transcutaneous electrical nerve stimulation) machines are used for chronic pain control and muscle stimulation. They work by passing short pulses of electrical energy though parts of the body. Not recommended if you are electrically sensitive."

Power Tools

The problem here is much the same as with devices like hair dryers and electric razors. If you're a professional, using high-power versions all day, it's more of concern than if you're just doing the occasional odd job around the house. It may be possible to find some models with shielded power cords.

Obviously, if you're pregnant, electro-sensitive, recovering from cancer or otherwise immune-compromised, you may want to consider your use of these.

Electrical – Wiring and Fuse Boxes

The Concern:

This was a surprise to me. Even when plugged-in things are not turned on, they emit an electrical field. This includes the miles of wiring in buildings.

You can test these lower frequency exposures, from the first window of our Spectrum, with a reliable electric field meter. It is worth it to have a professional testing of your home, office and school for these fields as the radiation can 'pool' in spots and the professionals know where, and how, to test for these fields. And their instruments are highly sensitive and reliable.

An EMR-qualified electrician can advise you and do the work on your house wiring – for instance installing demand switches to turn off circuits when they're not being used. Then you can retest on your own to assess the effectiveness of your interventions. This is certainly something to keep in mind when you're building a new house or about to do a major renovation.

(KG) "Demand switches should only be installed if proper testing was done first to identify the contributing circuits. To my knowledge, there is no consumer electric-field probe that would give you reliable readings. Body voltage testing can give you false readings, depending on the electrical conductivity of surrounding building materials e.g. in close range to grounding pads, stucco on wire mesh, metal roofs, etc."

Electrical Wiring – Safer Solutions

Building or renovating?

(CA) "Ideally, it is preferable and far less expensive and problematic to have an EMR professional create a low EMR wiring design and implementation – before the construction stage – using shielded cable throughout the building. This eliminates all electric fields from wiring sources."

Fuse Boxes – Safer Solutions:

(RM) "Locate household power meters and fuse boxes away from main living areas. They may be a source of high magnetic fields."

Power Cords

(RM) "Do not have power cords near your feet, or under your bed. Where power cords are necessary, use shielded ones as they reduce the EMF levels drastically."

Power Lines

If you live close to power lines, you should test inside, and outside, your house and property to know how much radiation you're being exposed to. Best to hire a professional, or get a Gauss meter and do it yourself.

Experts warn that these high-voltage power lines should not be close to schools, hospitals, or where people live.

Environmental health consultant Stan Hartman tells us:

> People should also know that neighborhood power lines are a concern, as they can have stronger fields than huge towering transmission lines. Be sure to check the lines at different times of day and on different days.

Here is an example of what we are up against. In British Columbia, the official response to Moms Against Power Lines, from a chief medical health officer, was "I've been following this thing for my 35 years in public health and there is no real danger."

Power Lines – Safer Solutions:

Ideally, you choose a neighborhood after you've checked out the radiation levels emitted from cell phone base stations, or radio or ham broadcast antennas, radar, or some power lines. Stan Hartman suggests the only solution is to paint all the walls and doors with shielding paint, and cover the windows with metal screening or metallic sunshield or shielding fabric. There's more about these products and associated cautions later in this section.

(KG) "If properly installed, metal screenings/shielding fabric could protect from the electric field exposure (and potentially from wireless transmitters if specified for it). One of the major components of power line emissions is magnetic fields, which the shielding paints and metal screenings would not address. The safest solution for power lines is to keep sufficient distance."

Vehicles

Electric/Hybrid Cars

Unfortunately, the energy-saving strategy of hybrid cars may have significant risks for human health. A hybrid vehicle combines a regular internal combustion engine and an electric motor powered by batteries. When you start the car, and at low speeds, the car runs just with the electric motor and battery. At high speeds the gasoline engine kicks in. The electric motor and batteries help the conventional engine work more efficiently, so far less fuel is used compared to even very efficient gas-powered vehicles. A good goal.

Electric/Hybrid Cars – The Concern:

However, these vehicles have big battery packs just behind the rear seat, with cables running underneath the passengers toward the front of the car. Some drivers have used Gauss meters to check inside their hybrid vehicles for the EMR levels from the flow of electrical current to the motor. They reported very high levels – hundreds of times higher than our scientists consider safe.

There are also anecdotal reports of drivers of hybrids complaining of new health problems including rising blood pressure, excessive fatigue while driving, and other symptoms of electro-hypersensitivity. I hear they are developing hybrid trucks and buses – very big batteries. Good for global warming, but what about the health of the occupants?

(LG) "The magnetic fields are approximately 10 times higher in hybrids than in conventional vehicles, depending on the measurement location in the car."

The back seat reading can be extremely high. Is this really where you want to put your baby's car seat?

Hybrid Cars – Safer Solutions:

Shielding for the batteries. Unlike in mobile communications, there are no information-carrying radio waves here, so the batteries can be completely covered with a special metal that blocks the EMR, without disrupting the efficiency of the battery.

(KG) "In order to block the magnetic fields from the batteries, the current-carrying wires would also have to be shielded."

EMR in Your Gasoline-Powered Car?

Even in the non-hybrid vehicles, there are hundreds of electrical devices producing strong fields. Our car has electrically-heated seats. I used to think this was a great winter feature.

The Concern:

(KG) "Most cars emit substantial amounts of magnetic fields. Depending on the tires and the location of electronics and wiring, exposure levels can vary greatly in conventional cars, and can be just as bad as in hybrid cars."

(AP) "Cars have electrical and electronic equipment (power wiring, fan motors, computerised controls and dashboards) that can disturb electrically sensitive people. The front seat can be a particular risky area.

Some upmarket cars have electronic control units under (or even as a part of) the driver's seat. These will give off high magnetic fields. Also worth checking is whether the angle and position of car seats are electrically controlled. These have higher EMF/EMR than mechanically operated seats."

(LG) "Other features in conventional cars can cause discomfort e.g. an air conditioning system control that sits in the ceiling right above the driver's head and puts out a high magnetic field. This has given some a blinding headache. When the cause was discovered and the equipment disconnected, the headache never returned."

(CA) "In gas-fuelled vehicles, spark plugs firing repeatedly a few feet from occupants causes RF exposure. Many EMR-sensitive people react to this. Diesel cars without spark plugs do not have this problem."

Newer conventional vehicles also present more exposure than the older cars do, because of the on-board computers, and other high-tech additions.

Non-hybrid Cars – Safer Solutions:

If you're considering buying a new vehicle, or when you're booking a rental, it's preferable to get one with as few electronic bells and whistles as possible. Choose your car carefully, preferably using a meter (see Resources at the end of the book) to detect the fields. And, to be mindful of our environment, you can always opt for the most fuel-efficient conventional vehicles, limit your amount of driving, limit speeding, carpool etc.

Global Positioning Systems (GPS)

Have you been reading this and wondering if there's going to be any GOOD news? Well, here's some. Since the first edition of this book, I have learned that the on-board GPS is *not* a concern.

(LG) "The GPS is a receiving-only device emitting a small amount of radiation – more like a radio."

However we are being exposed to more and more wireless radiation as we travel.

Wireless on the road and by rail

Someone contacted me recently to ask about wireless access in vehicles (via Bluetooth-like systems). Evidently, some new cars have this built right into the sound systems, but are activated with the ignition so you do not have the option to drive the car without being exposed to this kind of radiation. A confined metal space, experts advise us, can increase the harmful effects of EMR.

One of my neighbors was horrified when I told her about this:

You're not serious? There was just a huge study showing wireless is safe. Our new high-tech car is outfitted with wireless access. The kids think this is very cool.

We even got them wireless headphones.

I thought we were doing the right thing, and that the onboard Wi-Fi and DVD player were going to be so handy. This is frustrating. How are we supposed to know?

If you have traveled lately by train, or a long-distance bus, you know that many are now "wireless hot spots" whether you want it or not. The health effects are just one of the concerns; as you will see in the next section, this kind of radiation has been shown to affect cognitive abilities including reaction time.

Having the conductor, or driver, texting is an obvious distraction, but making transportation vehicles wireless presents other hazards. This is particularly worrying in a high-speed train, or in an aircraft.

Airplane Travel - Wireless In The Sky?

Alasdair Philips, the electrical engineer heading Powerwatch, has done onboard testing of the electro-magnetic fields in aircraft cabins.

He warns:

> Some airlines are planning in-flight mobile phone and Wi-Fi service. They are installing a picocell base station – effectively, a small mobile phone mast – onboard the aircraft. Passengers web-browsing and video streaming greatly increase the RF levels in the cabin.

Many experts are seriously alarmed by this. A group of European physicians, the German Environmental Physician Initiative, headed by Dr. Scheiner, wrote a letter (the complete letter is in Resources) to many airline companies about how in-flight wireless can affect the passengers, and the crew – including the crew's cognitive functions. Now that's a scary thought.

Yes, the proponents of this technology have tested the effect on the aircraft navigation systems. No problem, evidently. Have they, however, tested the impact on the pilot and other air crew?

Radiation from Mobile Phones and Reaction Time

Dr. Hans Scheiner:

> I have talked with a pilot who explained that the technical problems – potential interference with onboard navigational equipment – have been solved. However, there's a biological problem to consider. There are not yet any scientific studies of the biological effects on the crew and passengers, but there is enough evidence to cause concern.
>
> While the airplane is moving at an altitude between 8,000 and 12,000 metres, a reduced air pressure occurs inside of the plane, which equals the air pressure of 2,000 – 3,000 metres outside. Therefore the breakage of the blood-brain barrier is more likely because of the lack of oxygen and the well known altitude sickness.
>
> The severe consequences of the brain and nerve damages and safety of passengers are very concerning especially those of the pilots, because they are already highly exposed from radar. The symptoms caused by high frequencies like headaches, drowsiness, vertigo, nausea are often connected with loss of hearing and vision; lack of concentration and memorization are in this context known as the "Microwave-Syndrome" (Johnson-Liakouris, 1998, Mild 1998, Santini 2001, 2002, 2003, Navarro, Oberfeld 2003).

> A British lab found that reaction time is much longer when someone is exposed to the high frequencies from Wi-Fi and mobile phones.
>
> This is particularly a concern with the pilot's reaction time, when these high frequencies open the blood-brain barrier. This could become a cause of an aviation accident.
>
> Some Canadian researchers at the University of Toronto also found that drivers of vehicles were nearly five times more likely to have an accident when people were using mobile phones in the car.

Remember the list of EMR exposure symptoms: difficulty focusing, concentrating, and staying on task. Cardiac irregularities, unexplained anxiety, and other symptoms are never good at 37 thousand feet. More air rage incidents?

And remember the experts cautioning us that radiation is greatly magnified in a confined metal space, one that is already high exposure.

Alasdair Philips reveals:

> Planes are not an EMF friendly environment even now. Most have in-flight video screens built into the back of each seat. These are a source of high-frequency fields both for the people watching the screen and for those whose seat it is fitted into.
>
> The personal lights above the seats in a plane are often high-frequency fluorescent lights and can be sources of significant electric fields. Turn yours off.

With added wireless Internet, the RF radiation will bounce up and down the cabin 'metal tube'. There are also signals from various plane transponders mounted under the fuselage which make their way up into the cabin at surprisingly high levels.

Many planes also have an emergency Iridium phone system that is internally active all the time they are in the air, with active handsets at both ends of the plane using a pulsing DECT (digital) cordless phone like signal 24-7.

There was a tragic story of a co-pilot who had a mental breakdown over the mid-Atlantic, and was carried off the plane yelling, "You can just email me". This unfortunate incident could have nothing to do with radiation exposure.

Yes, people are probably clamoring to use their personal electronics onboard. Airline marketing departments take note: Don't you think that most people value being safe in the sky more than convenience?

Airline litigation departments: think of the ramifications if there's an incident attributed to this, and it is proven that no one pre-tested the acute, or long-term, effects of this radiation on the cognitive functions of pilots. And airline companies ignored medical warnings.

I used to work in the aviation industry doing programs for air traffic controllers and pilots, and I know what kind of stress they are under without this added load. I was also a member of the Canadian Aviation Tribunal.

I predict we will see a reversal in this policy when airlines realize the risks, and when passengers and the well organized pilots' unions are better informed.

We need preventive action, not an incident.

WASHINGTON, Oct 26, 2009 (Reuters) – "Pilots of a jetliner that overshot its destination by 150 miles last week told U.S. investigators they became distracted during an extended discussion of crew scheduling that included their use of personal laptops."

Evidently, their airline 'offers wireless Internet on most flights'. Having read what our experts have suggested about how this

radiation can affect cognitive abilities, and remembering how microwaves ricochet in an enclosed metal space, are you also wondering what effect the wireless system might have on a flight crew's gap of attention?

If you know any pilots, flight attendants, air traffic controllers, or airline company executives, please alert them to this potentially disastrous wireless wake-up call. I am trying my best to get the word out but the company's answer is usually, "The passengers are demanding this onboard convenience."

Remember the children holding the sign, "Is it worth the risk?"

I have worked with many pilot groups and have great respect for their professionalism. No one cares more about flight safety than they do. We should listen more to their concerns.

Here's a quote by "Ask The Pilot" from salon.com:

> Cockpit hardware and software use radio transmissions for a number of tasks. Whether transmitting, receiving or simply sitting idle, cell phones are able to garble these signals. As you might expect, aircraft electronics are designed and shielded with this interference in mind. This should mitigate any ill effects, and to date there are no proven cases where a cell phone has adversely affected the outcome of a flight. But you never know, and in some situations – for instance, in the presence of old or faulty shielding – it's possible that a telephone could bring about some sort of anomaly.

Anomaly?

Security is also an issue, advises a computer consultant, as the passengers could well include hackers with nothing to do except poke around your files. There are also concerns that online connectivity in the skies could facilitate communication between terrorists planning, or co-ordinating, an attack.

And there are privacy issues. Who wants loud, endless chatter, or worse, on nearby mobile devices.

If you are electro-sensitive it may be difficult already for you to travel. As more of us become electro-sensitive, we will need special areas set aside in airports that are "Wireless Free".

Safer Solutions? Let's lobby the airlines, and regulatory agencies, to test the biological effects of accumulative EMR exposure on flight crews, and passengers, not just the technical equipment.

Another Aviation and Wireless Issue?

In 2003 the SARS virus, and in 2009 the swine flu epidemic, sped around the world, often by air travel, threatening a global pandemic. As wireless radiation has been shown to impair the immune system, we must consider the impact of millions of people traveling in wireless-enabled planes, and trains – confined metal spaces – during these outbreaks.

Hotel/Motel Accommodations

Medical spas, yoga centers and roadside motels are advertising "free wireless Internet". As more of us request wireless-free accommodations, this will change. Look for places that at least restrict the wireless to public spaces, and/or provide hard-wired access in the sleeping rooms. You will sleep better, think more clearly, and be more relaxed, the experts advise us, if your room is not in a wireless zone. It's also wise to unplug cordless phones, or at least move them away from the sleeping area. Most places have corded phones for power outages.

Radiation Rescue Retreats™

We are going to work with places that can provide a respite from EMR exposure. This will bring several benefits. Being in a low EMR environment, even for a short time, will give us a radiation break, and help determine if symptoms and/or conditions may be related. While we retreat and recharge, we can discuss the challenges of going lower tech, and support each other in this journey. Check our website radiationrescue.org for details.

These retreats will take place in wireless-free family campgrounds, yoga/meditation centres and resorts – from rustic to up-scale luxury. One of the centres where we are planning these retreats is a medical clinic that is already an example of a well planned low EMR building – Sanoviv Medical Institute, in Baja California. Every possible way to reduce electro-magnetic radiation was incorporated into the structure of the buildings during the planning and construction stages. All electrical wiring is specially shielded, lighting in guest areas is low voltage, and electrical outlets and devices are located as far as possible from guestroom beds. Radiation-emitting devices such as cell phones and laptop computers are also restricted from guests' rooms. The rationale behind these precautions is that Sanoviv guests are likely to achieve healing more rapidly if they are not having to deal with toxic electro-magnetic pollution during their convalescence.

Safer Technology?

Shielding electrical wiring, better lighting etc. are a few ways to reduce our EMR exposure. Many of our contributing experts are also calling out for the development of safer – less harmful – devices. There could be safer cell phones and, yes, even safer wireless networks.

Building Biology experts W. Maes and F. Mehlis have raised this vital question in *Wireless Network, A Flyer by the Building Biology Association (VB), 2009:*

> Why are such antennas not activated except when they are actually needed, e.g. for the short period of time during a data transfer, and then automatically deactivated as soon as not in use? This would have been easy to accomplish, but the technical developers did not consider the possibility. An automatic deactivation was deliberately shortchanged.

Other Ways To Reduce Our EMR Exposure?

With so many of our younger generation using mobile phones, wouldn't it be comforting to think that there was a quick and easy solution to protect them from the radiation? At this point, there does not seem to be a scientifically sound way to do this, though there are lots of products out there making big claims.

Chips, Pendants, Jewelry etc.

People often ask me about the numerous devices on the market, such as radiation-deflecting/harmonizing chips that you stick on your cell phone or laptop, or about jewelry – pendants or bracelets – that you wear.

These kinds of interventions are described as enhancing your own energy field in some way, and/or harmonizing/neutralizing the artificial electro-magnetic fields.

Some of these products seem to be helpful, for some people, at least for a while; many do not. I tried some of these myself, at one time. This technology, while promising, does not seem to provide the answer, just yet.

Before investing, and trusting in this protection, you need to see some solid science on their effectiveness – not just testimonials. And insist on seeing the scientific evidence of how long the reported protection lasts. Some of the most popular brands were tested by an independent laboratory, which reported the effectiveness wore off after three months. If this is true, are the consumers aware of this?

And some clinicians advise that for some electro-sensitive people, these kind of protective interventions are contra-indicated because they may make symptoms worse.

The safety consultants I've talked with are concerned that 'magic bullets' like chips and pendants may lull people into a false sense of safety – they keep chatting away on their cell phones, keep their cordless phones and wireless router radiating day and night, believing they are now safe. Seems more research is needed.

Sensory Perspective in the UK has done an excellent service offering people the information and tools to protect themselves. The organization does not, however, recommend chips or other similar interventions.

One of their technical directors explains:

> Sensory Perspective focuses on what it knows will work: certifiably calibrated measuring instruments for detection, EMF barrier products for protection. We simply suggest that one uses suitable barrier products (shielded curtain materials, wallpapers, paints) for the home or office. www.detect-protect.com

Physical Radiation Barriers

These kinds of products are supposed to work by deflecting/blocking the radiation in some way. Stan Hartman comments on one example, pocket protectors: "I haven't tested any of them, but assuming that they shield the high frequencies from the phone, that still leaves the low-frequency magnetic fields from the battery-saver technology, which they can't shield. It's best to just keep the phone at a distance in a purse or backpack or something – or toss it, of course. Certainly don't put it in a pocket over your heart."

Shielding Materials

If you've done everything you can to reduce your exposure, you may still want or need more protection from sources you can't control. In that case, you might have to do an EMR-repelling reno. This might include special window screens or films, wallpapers and EMR absorbing paints.

Stan Hartman gives us some expert advice on shielding:

> Keeping out microwaves is a little like trying to keep out blowing sand, though – you can do it, but you have to be sure there aren't any sources inside or any crevices for the radiation to get in through.

That means no CFLs or other fluorescent lights, for instance, or CRTs – the old TV sets and computer monitors – or microwave ovens. If you have inescapable internal radiation sources, you could be better off not shielding at all and just trying to avoid radiation while you're sleeping at least, with a bed shield.

When it comes to shielding fabrics, wallpapers, paints, etc. caution is required. Shielding curtains, and bed canopies, are made from a fabric made of RF reflective fibers. This fabric is very effective at blocking RF – but you MUST use it properly, or you can get badly zapped! Some of these products have to be grounded. Do not use any of these without professional consultation, even if it's just a phone conversation with the supplier.

For example, the bed canopy fabric has to be properly draped away from the edges of the bed so it does not touch you, or any pet sleeping nearby. You cannot have any EMF/EMR sources inside the space draped by the fabric – no regular lamps, plugged in clocks, alarms etc – nothing at all wired or wireless.

So how do you read? A non-metal lamp, with an incandescent light bulb (not halogen or fluorescent) and shielded power cord is recommended.

And you need demand switches to reduce the electric and magnetic fields in the room before you use the fabric. A demand switch is a smart electronic device that automatically cuts off electricity when there is no demand for it. It also supplies electricity when there is a demand. Saves on your power bill, as well. This is one area where, unless you're a licensed electrician, you'll need professional help.

It's important to note that shielding devices, and therapeutic interventions, that may be very helpful for people without symptoms, can be devastating for electro-sensitive people, if not used properly.

At the end of the book, in Resources, I have listed sites in several countries, operated by qualified experts. You could begin your research there if you're considering adding these to your home.

Where To Start?

As you contemplate changes you might want to make in your family's technology habits, this thoughtful message from our "library mum" may be useful:

> As you know, my life is very hectic but I am now taking a break more often to get outside with the kids. It's great for them, and for me. I'm not even texting at the beach.

And, thanks for Chapter 3, it was lucid, well set out and not frightening, after all. After reading it last night, I became convinced that it is necessary for us to have the discipline to do something about our cell phone use.

As you know, we (my husband and I) run a couple of restaurants. When we started, we worked all day together. After my dizzy spells, I had to go off of the front line and I am now rarely at the restaurant at all. Since I've gone home, we use the cellphones A LOT – mostly because we miss each other and also to make business decisions, ordering etc. In addition, I've always been "on call" to ensure that suppliers, or our front line staff, could reach me at any time.

Last night I proposed that we get pagers. My husband disagreed, saying that if he is going to reduce the cell phone exposure, why would he replace it with exposure by some other gadget? It was a good argument. Especially, since this whole reduction of EMR has to do with simplifying life.

So with these thoughts, this morning, we turned my husband's cell phone off, but he still has it with him in case he needs to make a call. My cell phone is on today, but it is hidden in a location away from our house. It is on so that I can put it on call forwarding. I will make a mental note to see if our server can do this automatically so that I don't have to have my cell phone on at all.

On the weekend, I was talking with my neighbour who refuses to use cell phones but is reliant upon her mobile indoor phone. We were talking about how much we hate to be tethered to a corded phone... that the mobile phone enables us to multi-task ie. doing laundry, washing dishes etc. while talking.

There was a small silence and then we both expressed a similar thought... maybe we should be paying full attention to the person on the other line? Maybe we should be visiting in person? Maybe the technology has been leading to a cheapening of human relationships?

There is a very cute part in the movie "Mall Cop" when a young girl is asked by an assailant to give up her cell phone, saying, "You must have one, all young kids these days have cell phones", to which she replies, "I prefer written sentiments."

Have you noticed how much more a handwritten letter or note means these days?

So, I see *Radiation Rescue* can be read on various levels. It can lead to basic changes – but it can also lead to quite profound changes. It has already instigated these more philosophical reflections on our life choices.

Now that you understand – from Step 1 – why so many experts are concerned, and know how to recognize the related symptoms and conditions – from Step 2, and have learned what EMR sources you need to address – here in Step 3, you are well-outfitted to draft your priorities. There is a detailed Action Plan at the end of Step 4.

Our Radiation Rescue Intentions

Top Priorities - what I want to tackle first:

_____

_____

_____

Secondary priorities – what I want to deal with after that:

_____

_____

_____

I should consult a professional to assess the following:

_____

_____

Other environmental exposures I want to address:

_____

Step 4. Enjoy Your Family Action Plan

Is anyone in your family already experiencing a lot of possibly related symptoms – difficulty sleeping, dizziness? Or one of the conditions that may be connected – headaches, insomnia, learning disorders, autism/Asperger's etc.?

If so, you will want to approach this step more vigorously as these people are more vulnerable.

If no one is experiencing symptoms, it does not mean the radiation is not having biological effects. It does give you the opportunity to reduce your family's exposure and risk of these symptoms and conditions.

You will also want to keep in mind the factors like proximity, intensity, duration/frequency as you decide on your priorities for action.

Room by Room – Audit & Recommendations

Fairly soon in my wake-up call I decided not to bog down in guilt, resentment or denial. I became determined to rescue my family and sought help, as I have explained, from the best sources I could find. Now all of that is set out here for you. Keep in mind this is not a quick fix; it is a process that takes some time.

My family began with a professional assessment. I strongly urge you to do the same, particularly if you are considering shielding fabrics, or other interventions that can cause problems if not used or installed correctly.

In the case of retro-fitting your home, or anything to do with electrical wiring, or shielding technology, it's wise to consult with and hire qualified professionals. Some of these are listed in the Resources section.

Testing of Your Home, Office and School

The Do-It Yourself Assessment – consumer devices to measure EMR are very helpful, not only in your own environment – but also vital when traveling (see Resources). You must know the basics, however, about the instruments, and what you are testing, or you can make your situation worse, not better. Like my friend with the microwave oven.

The Professional Assessment – this is key, in my view. Here is what Rob Metzinger (RM) advises: "Before the proper solutions can be implemented the primary sources of the problem – internal and external – must be identified."

Testing of Electric Fields, Magnetic Fields and RF (radio frequency) should be done with the most sensitive testing equipment available.

The professionals' recommendations should look something like this:

1. Devices/sources to remove, perhaps in an order of priority for your situation: the urgent first-to-go, and then remove as you can, etc.

2. Devices/sources to replace, and with what ....

3. Alterations in position, use of the devices/sources etc.

4. Protective Interventions – shielding fabric etc.

Once solutions have been put in place the site must be re-tested to verify the efficacy of the products. Periodically the site should be retested to ensure the interventions are still performing and that new sources of contamination have not arisen.

Some Tips on Having a Professional Assessment

Before you have the pros come in it is helpful to label your electrical control panel. Make sure the person turning the breakers on and off is wearing non-conductive gloves, and is <u>not</u> electro-sensitive. Mark each light switch and wall socket (in pencil) with the circuit number, and the same number, obviously, on the control panel (in pen). The technicians will check this but it saves you money if you can get a head start on this step.

Then they will go room by room – and outside of the building – to test for the electric, magnetic and radio frequency.

As we move along through your house, in the following section, I'll share some of the things I learned during our audit, and add the experts' recommendations for safer/less harmful alternatives in square brackets, like this [...]

EMR Testing Equipment

If you are using testing equipment before, or instead of, a professional assessment, please be very careful.

1. Do you know how to use the meter?

2. Is it reliable – good enough quality? Consumer models may not measure the high frequency of that device – your cordless phone, for example, and give you a false 'safe' reading.

3. Are you using it for the right exposure? For example, you need a RF (radio frequency) meter to test the radiation from a microwave oven, cordless phone – the handset or base, cell phone, wirelessly-connected computer etc. Remember the middle window of the spectrum?

For electrical outlets, appliances, heating pads and other wired devices you need a Gauss meter. See the Resources Section for a list of meters. Refer to the Building Biology Guidelines in Step 1.

If you get readings much higher than those levels, I really encourage you to get a professional assessment done. This is crucial if someone in your family is ill, pregnant, very young or very old.

Refer to the Resources section for qualified EMR testing and remediation consultants. Most electricians are not experienced in this specific work and may assure you "there's no problem, it's all to code and within government standards." This response is a good signal to look for someone trained in this area.

Front Porch

Let's start this house tour right at the front door step. First, is there a wireless home security system, or a wireless front door bell [hard-wired options]. Is the front porch light fluorescent [low wattage incandescent]. An automatic garage door opener communicates wirelessly but is not considered a significant risk. [Just be mindful of this and don't point the clicker directly at anyone standing between it and the door. Same recommendation for the TV channel changer.]

Bedrooms

The sleep zone is crucial to your wellbeing. As you may remember, EMR is known to disrupt sleep in many ways. Is your bedroom a radiation hot spot, or a calm low-tech oasis?

Besides looking at your own sleeping area, check out the places where children and/or pets sleep.

You are looking for: everything wireless, everything electric – all things plugged in – including clocks, radios, portable phones, TVs, electric beds, heaters, heating pads, electric cords, battery chargers, wall sockets, fluorescent (including compact fluorescent) or low voltage halogen lights, dimmer switches, lamps with transformers or the clap-on, or touch-on feature, everything metal – lamps, bed frames, headboard, footboard, side rails and box spring mattresses. No, you don't need to toss all of these, just note them and identify the ones that are creating high EMR.

[To reduce your EMR exposure, remove or unplug electronic devices located in the sleeping areas such as: televisions, DVD/VCRs, electric clock radios/alarm clocks, stereo systems, heating pads, electric blankets, and waterbeds. Battery-operated alarm clocks are a good alternative, or at least put the electric one a minimum of six feet away from your head.]

[If you absolutely have to have a television or a computer – with hard-wired Internet – in your bedroom, put them as far away as possible from the bed, and put them on a power bar that you turn off at night – or on a timer power bar.]

[Replace an electric blanket or heating pad with a hot water bottle or gel pack, or use the electrical kind just to heat up your bed, then turn off and unplug the cord before you get in.]

Metal is not recommended in sleeping areas as it reflects, and re-radiates, radiation from all surrounding sources, worsening exposures. Like most of us, you probably need to budget for many changes here. Science tells us that children are at least ten times more vulnerable to any toxic exposure than adults, so they need more protection. [Making our children's bedrooms a priority, our technical experts urge us to: replace box springs with ones that contain no metal; replace coil and spring mattresses with foam mattresses made from low emission natural latex, or cotton mattresses such as futons (and to use natural cotton fabrics wherever possible to reduce static electricity); and to replace metal bed frames, metal cribs, metal bedside tables and lamps, metal bunk beds – with safer non-metallic types.]

[As we could afford it, I replaced these items, and our metal-framed bed with a solid wood one that even has wooden slats to hold the natural fiber futon mattresses.]

[My husband and I built a wooden reading lamp with a specially insulated electrical cord. We removed the halogen lamps with the transformers that had set the detector screaming.

And we replaced all fluorescent lights with regular incandescent ones.] Everyone in our family is mindful of energy consumption and we have many other ways to conserve energy and costs.

Ideally, and especially if you are suffering from any symptoms, shut off power in the rooms where you sleep and all the adjacent rooms, above and below this area, at night. This can be done by turning off the corresponding breakers, or by using a demand switch.

You may find inspiration in creating a low-tech, low EMR sanctuary where you sleep. Instead of a TV area, why not think about a space for calming practices like meditation, yoga or tai chi?

These have many benefits including balancing our life force energies that are so disrupted by electro-pollution – more on this later.

The Children's Room

You want to go through your children's room with a sharp eye of awareness. Check for everything that is plugged in, the lights, electronic equipment etc. [Move the beds/cribs away from the wall – from any wall sockets and even the wiring in the wall. Avoid the use of sensor pads under baby.]

I'm presuming you are onboard by now in removing everything that hints of wireless – Internet access, cordless phones, wireless baby monitors [replace with low-powered plug-in ones, if you must have one].

[Use wooden beds without metal box springs or metal coiled mattresses. Use incandescent lights and battery alarm clocks.] Remember what Alasdair Philips told us about night lights – it's best to sleep in as dark a room as possible and that [plug-in red or orange, soft glow ones are preferred over white or blue lights, if we must have one.]

You might discuss this lower-tech reno with your children before yanking out their electronic entertainments and explain why you want their bedrooms healthier and safer for them. As you know, these issues are not high on any child's priority list.

Bathroom

You might test the electric toothbrush, electric razor and the hair dryer – using a Gauss meter. And check for microwave radiation around any mirrors or windows using a RF meter.

I realized the electric in-floor heating was not a great idea. Turning the dial down to "off" is not enough. [I had it disconnected.] Is the floor colder in the winter? Yes. But with a thick, cotton bath mat, it's fine. Also during that reno, I had found a quite large mirrored wall unit that we hung above the sink. It seemed like a great idea at the time. When the radiation detector "ghost busters" did our home assessment, the radiation bouncing off that mirror and metal cabinet was extreme.

(We may have to rethink some Feng Shui principles; while mirrors are considered good reflectors of energy, they do however also reflect what we could call the "negative energy" of EMR). [Now we have a small wooden framed mirror, and a small wooden shelf on the side wall.] After this EMR-busting retrofit, the radiation levels in the bathroom are now minimal.

As we've discussed, the health hazards in our homes are not all from EMR. Plastic shower curtains were found to off-gas toxic chemicals – and were still off-gassing a month after the package was opened! [I also replaced the plastic shower curtain with a fabric one.] And you may know about the chlorine and other chemicals in our tap water. [We have a filter on our shower head and another in the kitchen for our drinking and cooking water.]

You have the option of chemically-scented personal care products, cosmetics etc. [or the less toxic alternatives]. I discovered that virtually all standard brands of deodorant contain aluminum. Just 'trace amounts'. [All brands in natural health food stores are labeled "aluminum free".

Same recommendation goes for baby powder and baking powder – check for aluminum-free brands.] Dousing a newborn with sweet-smelling powder laced with aluminum? What are we thinking?

One of my neighbours reports on her home testing

"On the weekend I used a device that measures wireless radiation. I was surprised, and concerned, to get elevated readings in the vicinity of my bed. At the time there were no obvious microwave sources that might provide this measurement. My cell phone and laptop were turned off and the wireless router was unplugged.

My bed is only a few feet from a large window – so it seems like the gauge was picking up microwaves from sources outside – perhaps neighbours' wireless systems, or local cell towers. It concerns me that I am being subjected to these elevated exposures overnight while I sleep.

The biggest shock, however, came when I measured the microwaves coming from our portable phone. Sitting at idle, the phone in its receiver gave a slightly elevated reading. But when I picked up the phone and stepped a few feet away from the base station and turned the phone on, the reading on the gauge spiked up to a really high 175! I was shocked.

I had been happily using the portable digital phone as an alternative to using my cell phone – thinking that my exposure would be markedly less. I never anticipated that the microwave emissions from this portable phone would be in the same range as my cell phone. In fact they were about 10 percent higher.

The take away message for me is to get a corded landline as soon as possible.

The three "hot spots" in my home were the portable phone, the cell phone and the wireless network router, so we can get our email in the kitchen.

I was surprised to find elevated readings coming from my cell phone – even when it was turned off. Suffice it to say, I am no longer using it as an alarm clock. It no longer has a place overnight on my bedside table. And we have switched back to hard-wired Internet access. We decided to make the bedroom a low electronic environment. And it is a great improvement."

Kitchen

Adults without health problems are fine in most kitchens, as long as they stand far away from microwave ovens when they are on. Those who should use more caution include: pregnant women, children, people with poor immune systems, convalescents, women with a history of breast cancer and men with a family history of testicular or prostate cancer.

When you're in the kitchen, you're often in the soup – of EMR. Test all plugged in appliances with a Gauss meter (microwave ovens also need testing with the RF meter, as you may remember). If you're shopping for new appliances, take the testing equipment to help you choose the best options.

How many counter-top gadgets do you have? [When I use the kettle or blender I stand back and unplug them when I've finished].

[When my fridge expired I replaced it with a no frills low energy consumption model. Each extra option that requires more power is going to increase the electro-magnetic field. I also removed the under-the-counter CD player/radio that was only a few feet from my head when I was in the kitchen. Low intensity but high proximity and duration. The same player placed on the other side of the kitchen is no longer a concern].

The dishwasher, as you know, uses lots of energy and water, so being concerned about these, we don't run it more than we have to, always use the quick wash/air dry cycle, and don't stand against it when it's on.

In our kitchen there's no microwave oven, [reheating food on the stove only takes a few minutes]. What kind of stove do you have? [Whenever possible I cook on the back burners on the stove as the electro-magnetic field from the front burners is much higher. The next time I buy a stove, I will get a gas stove and make sure it is well vented].

Gas stoves do not pose an EMR hazard as they don't use electricity. The sparks used to light the gas produce short bursts of high-voltage electricity, but these are insignificant. Some people with multiple chemical sensitivity need to avoid the gas fumes.

Ovens, including fan-assisted convection ovens, electric fans/vents, even smaller toaster ovens are generating strong fields. If you are standing close to the stove, with the front burners blasting, and the over-the-stove fan is on at your head, you are in a fairly strong field - test with a Gauss meter. [Again the recommendation is stand back. Proximity is one of the best guidelines to keep in mind.]

Laundry Room

The same is true in the laundry room. Washing machines and dryers create strong electro-magnetic fields so it is best not to stand right beside them when they are on.

The other issue here is the toxic content of most laundry products unless they are one of the non-toxic natural brands. I once felt like leaving a Pilates class because most women had either washed their clothes in the standard detergent – chemically perfumed – or dried them with scented anti-static dryer sheets. I wanted to tell them that the chemicals from these products are absorbed through our skin. Ever walk past a house where people use these? You can detect the smell from the dryer exhaust.

Vacuum Cleaners

Most vacuum cleaners have powerful motors and give off high EMR. Strap-to-your-back kinds are *not* recommended. If you are pregnant, upright models are also not recommended. [The canister kind is preferred as the motor is further away from your body than upright models. Use a plain mop or broom for non-carpeted floors. Best not to use chemically-treated floor cleaning pads.]

Office

Whether we're talking about your home office, study, den or even your professional office, the same basic approach applies. We worked with the strongest exposures first – cordless and cell phones, other wireless devices, wireless Internet access and lighting. Then we tackled anything wired.

Your Cordless Phone Audit

How many cordless phones are in your home and office? Are they analog or digital? [We replaced all of ours with corded landlines that don't require electrical plugs, and use small batteries for the caller ID and speakerphone features. You can also replace the handset cord with a 25 foot one, for long range wandering.]

Wireless routers

Proximity is a factor, even with a long-range device such as this. [Every EMR expert strongly recommends replacing wireless Internet with a hard-wired system.

You might check out the kind that connects your Ethernet through the house wiring; you can plug into an electrical outlet for your Internet access. Not inexpensive, but has its benefits in reducing significantly your EMR exposure by taking wireless Internet out of your home.

If you must have a wireless router, place it as far away from sentient beings as possible, only power it on when you use it, and turn it off after that; always have it powered off at night.]

The radiation range from the average router is about 175 metres, (575 feet). At night, and when you're away, be considerate of your neighbors and switch it off. Any bees and other migrating beings whose natural navigation systems are being thrown out of whack by EMR will also be grateful.

It is recommended to stand back from all office equipment, printers, copiers, fax machines etc, when they are working.

And I won't mention another word about the microwave oven.

Experts' Action Plans

Many of these points have been touched on in previous sections, but it's interesting – and inspiring – to see how the leading authorities are putting their EMR-reduction advice into practice in their own homes and work places.

Dr. Martin Blank has no cell phone. His wife takes one in the car in case of an emergency. He has replaced his cordless phones with corded landlines. He's moved the electric clock away from the bed, got rid of circuits that have multiple electric switches on one line, as well as dimmer switches and instant-on devices.

He has also got rid of fluorescent bulbs in his house because they give off too much RF noise that has been linked to an increased risk of cancer.

Dr. Blank has meters and measures EMF levels at home from time to time. Instead of wireless Internet, Dr. Blank has a cable linkup between computers that are on opposite ends of the house. He and his wife do not own a microwave oven.

Dr. Hans Scheiner and his wife have had to make changes to their house because of their exposure to EMF from the proliferation of antennas in their area.

Testing with meters found the levels were very high in Dr. Scheiner's main floor medical clinic and in the 3rd floor bedroom. He has had the clinic walls coated with a special paint with small particles of carbon in it.

In the bedroom, the Scheiners have curtains made of shielding fabric, and on the outside wall have installed a thin metal plate that attracts electro-magnetic frequencies and grounds this energy into the earth. "The levels are much lower now."

Dr. William Rea offers his strategy for protecting his family from toxic exposures: "No cordless phones; no fluorescent lights, keep the chemical load down, consume organic food and safe water."

Rob Metzinger tells us:

"First, using special meters I evaluated exposure for Electric Fields, Magnetic Fields and RF (radio frequency from microwaves – neighbors' cordless phones, wireless networks etc.) in all areas of the home, concentrating on the sleeping areas. Turned off power in all sleeping areas reducing electric field exposure by using demand switches.

- Removed all electronics from the sleeping areas.

- Installed shielded power cords on all computers and electronics.

- Moved computers and monitors as far away as possible from the users.

- Moved air conditioner away from children's sleeping area.

- Installed shielding curtains to deflect high levels of RF from entering the home.

- Replaced low voltage halogen, fluorescent and compact fluorescent lights with incandescent bulbs.

- Replaced dimmer switches with regular on/off switches.

- Replaced spring mattresses with natural latex foam mattresses.

- Replaced electric alarm clocks with battery-operated units.

- Made a battery-operated night light for my children's rooms.

- Installed one analog 900 MHz cordless phone. All the other phones in our home are now land lines.

- Periodically retest exposure levels throughout the home.

- When purchasing any new electronic devices, I test them for EMF/EMR emissions."

Action Plan – Other Key Wellbeing Factors To Consider

Because we can't control everything around us, even the best action plan, up to this point, is reducing our chemical and electrical exposures, but not eliminating them entirely.

At the end of this section, you will see the detailed Action Plan to fill out, if this is what you want to do.

You or someone you know may already have some of the EMR-related symptoms and conditions listed in the questionnaire in Step 2. Even if you haven't, you'll benefit from the recommendations from some eminent complementary health practitioners on how to protect and improve your health in this hyper-tech world.

First, have you ever wondered how the health of people in our advantaged society got so far off course? Let's take a look at how this might happen. This analogy of a river gives us insights into our own health challenges, and wellbeing strategies.

The River of Health – pristine or polluted?

The River Runs Clear

The clear mountain stream flows unhindered – translucent, its foundation bed of smooth river rocks can easily be seen. Light sparkles on its pristine surface and if you were to drink this pure, oxygen-rich water it would taste fresh and almost alive.

It is mineral rich, being fed by pure mountain streams, so its acid alkaline balance is as nature intended. The rocks over which it travels are free of debris. There are no blockages in the river; it is not twisted and its flow is not blocked. The river banks are clean and free of clutter.

Then at some point toxic sludge enters the river – at first this is absorbed by the river, diluted and purified. Then at another place downstream, more sludge is introduced and bit-by-bit the river's ability to absorb and detoxify these toxins becomes compromised.

The once-clear waters become clouded; the river bed covered with a dark slime. Debris builds up on the river bank and the river narrows. In places, the river slows to a congested trickle. Problems arise at that blockage.

The flow of energy, blood and lymphatic fluid that sustain our health, and our lives, is much the same. Its path, twists and turns are partly set by our genetic heritage and family health history, but everything that enters it has an effect.

Ideally, we want to start our lives with as clean and strong a river of health as possible, but this is an increasing challenge in the modern world.

The average North American diet contributes a great deal of sludge to the river of life, including highly acidic foods (disrupting the pH balance), pesticides and herbicides, assorted additives, trans fats....

And emotional and/or physical traumas early in life also take a long-term toll on the body's ability to heal itself.

The river reacts

Most children consume vast amounts of sugar and packaged foods – weakening the immune system and leaving them vulnerable to infection. Some children get recurring infections and lots of antibiotics. This can become a worrying cycle.

Antibiotics are truly wonder drugs and can save lives. Even if the family physician is careful only to prescribe antibiotics when strictly necessary, children may be ingesting antibiotics from foods such as industrially-raised chicken. Antibiotics kill the bacteria we don't want, as well as wipe out the friendly bacteria – intestinal flora – that we need to maintain a healthy balance. A yeast overgrowth can result, overloading the immune system, leaving us vulnerable to more infections. If we are hit with a 'superbug' we may have nothing effective left in the arsenal.

Chemical sensitivities, asthma, even childhood cancers can arise from a compromised immune system, many experts advise. Systems are further burdened with mercury from vaccines and dental fillings.

The typical teenage diet of sugar, soda, fries, pizza and other fast foods is highly acidic. And many young people don't get near enough sleep.

We have now set the ground for the potential of a lifetime of illness and disease. Later on in this section we'll look at a number of factors that combine to support our overall health. We can turn this around.

First, we'll start with electro-sensitivity, since this is a new condition, and most people – even many family doctors – are not aware of how this exposure affects every system in our bodies.

Electro-hypersensitivity (EHS) – a new medical condition

Most of us are now aware that there are people who are very affected by environmental pollutants, whether these are man-made chemicals or naturally-occurring allergens such as molds or pollen.

People with Chronic Fatigue Syndrome and Fibromyalgia are finally being taken seriously instead of being told their symptoms are all in their heads. And now, we have a whole new syndrome that most of our doctors do not know about yet.

It's called electro-sensitivity (ES), or electro-hypersensitivity (EHS): a phenomenon in which individuals experience recurring health effects which flare up while, or just after, they are exposed to electro-magnetic radiation.

As Dr. Magda Havas points out, there are heavily-radiated parts of the world where 3% of the population is so severely affected they have to live as far as possible from wireless radiation, and wired electrical exposure. And she reports that as many as 35% have symptoms that allow them to function but may significantly impair their quality of life.

In Sweden, EHS is officially recognized as a disability, so employers and building owners are expected to try to rectify the problems and sufferers may be entitled to disability benefits.

While EHS is not an accepted medical diagnosis yet, the Canadian Human Rights Commission recognizes it as an environmental sensitivity and classes it as a disability.

Sarah Dacre, one of the leading advocates on this issue, was a television producer in the UK. She has helped many people with her articles on her journey with EHS. Here is an excerpt.

Allergic To The 21st Century – my battle against ill health caused by EMR, by Sarah Dacre:

> I am unable to live and work in a conventional way since being struck down with electro-sensitivity (ES) in May 2005. My health has been badly affected. When I travel I am affected by others using mobile phones, proximity to masts, emissions from Wi-Fi cafes, the list is endless. There are many of us affected. We are effectively allergic to 21st century living.

Dr. Scheiner has treated thousands of people with symptoms related to EMR exposure. He agrees the number of people with EHS and related conditions is on the rise:

> Some people are electro-allergic – not only sensitive but hyper-sensitive, to the point they have to become recluses and live in the country, where there are no antennas (cell towers). When they are exposed they have headaches, sleeping problems, exhaustion during the day, dizziness, vomiting, tachycardia, concentration and memory problems. They may faint or their vision might be impaired enough that they have to stop driving. It's likely these are signs of opening of the blood-brain barrier from exposure to EMR. Doctors often tell them their symptoms are psychological.

Dr. Scheiner says an electro-hypersensitive person has warning signs he or she is being affected, and can usually recover from the symptoms by avoiding the exposure.

However, the next step along the continuum is what he calls an electro-allergic reaction, in which the body's response is a real illness.

Some people, Dr. Scheiner reports, will feel no effects or symptoms of exposure – then suddenly will be hit with something as serious as cancer, high blood pressure or a stroke.

Thomas M. Rau, MD, the Medical Director of the Paracelsus Clinic in Lustmühle, Switzerland, is convinced electro-magnetic loads lead to cancer, concentration problems, ADD, tinnitus, migraines, insomnia, arrhythmia, Parkinson's Disease and even back pain:

> At the Paracelsus Clinic (www.paracelsus.ch), cancer patients are now routinely educated in electromagnetic field remediation strategies, and inspectors from the Geopathological Institute of Switzerland are sent to patients' homes to assess electro-magnetic field exposures.

Source: www.electromagnetichealth.org, 'Electromagnetic Load a Hidden Factor in Many Illnesses' February 2009

No, EMR does not cause everything

At this stage you may well be wondering how electro-magnetic radiation can be blamed for such a lengthy and diverse list of symptoms and conditions.

According to data compiled by Paul Raymond Doyon, MA, research indicates "there are about eighty immune system disorders that we didn't have twenty-five years ago before we really started microwaving the planet." Is there a connection, one wonders?

Symptoms may suddenly appear in people who have had a cell phone tower or a wireless hub installed near their home, or who have begun living or working around high-powered wireless, or wired, equipment.

Whether a person develops symptoms, or not, can be influenced by the intensity and duration of the exposure, and by the individual's vulnerability – whether his or her overall health and immunity are strong enough to buffer the effects of EMR.

Reportedly, that is why EHS is more common in people who already have chemical, or other environmental, sensitivities. Many may also be burdened by heavy metals in their systems and by food allergies.

Dr. Scheiner also says people who have had cancer are more susceptible to developing electro-hypersensitivity; their bodies are likely too busy fighting the cancer to be able to fend off this extra assault.

Even when exposure to EMR hasn't directly caused a disease, for instance diabetes, or a neurological condition such as MS, ALS or Parkinson's, the medical experts tell us that the symptoms may get worse as the person's body tries to cope with the additional burden of this exposure.

EHS Symptoms and Related Conditions

(with the kind permission of Susan Parsons: www.weepinitiative.org)

SYMPTOMS OF ELECTRO HYPERSENSITIVITY					
Neurological		Cardiac	Respiratory	Dermatological	Ophthalmologic
headaches	depression	palpitations	sinusitis	skin rash	deteriorating vision
difficulty concentrating	anxiety	pain or pressure in the chest	asthma	facial flushing	pain or burning in the eyes
muscle and joint pain	confusion and spatial disorientation	low or high blood pressure	bronchitis	itching	pressure in/behind the eyes
memory loss	fatigue	shortness of breath	pneumonia	burning	floaters
dizziness	weakness	arrhythmia			cataracts
nausea	tremors	slow or fast heart rate		swelling of face and neck	
irritability	muscle spasms				
numbness	leg/foot pain				
tingling	"Flu-like" symptoms				
hyperactivity	fever				
altered reflexes	insomnia				
OTHER					
digestive problems	abdominal pain	testicular/ovarian pain/swelling	enlarged thyroid	great thirst	dehydration
nosebleeds	internal bleeding	hair loss	pain in the teeth	deteriorating fillings	Light sensitivity
swollen lymph nodes	loss of appetite	hypoxia	allergies	frequent urination and incontinence	night sweats
immune abnormalities	redistribution of metals within the body	ringing in the ears (tinnitus or similar chronic ear-noise)	impaired sense of smell	altered sugar metabolism	dryness of lips, tongue, mouth, eyes
Severe reactions can include seizures, paralysis, psychosis, and stroke					

Step 4. Enjoy Your Family Action Plan

Doesn't it seem incredible that just about everything you've ever heard of is here? How could this be? Medical experts I've talked with explain this by saying that EMR affects every cell, organ and system in the body. So if you consider this, it doesn't seem as hard to believe that there are so many related symptoms and conditions.

If your house were flooded repeatedly, eventually everything within it would be affected.

How can you tell whether your symptoms are EMR related, or not? This is just what I asked Dr. Scheiner. He explained that each symptom might be unspecific (in other words not related to a particular illness):

> You can have a headache for a lot of reasons. It gets specific if you get lots of them, or sleeping problems. If you can't sleep at night this could be microwave syndrome from EMR exposure... not yet a sickness, it's a warning syndrome.
>
> The body is saying 'I am having trouble here'. It is a call for help. Allergies and food sensitivities could also be a test.

More women than men? Not so.

Dr. Scheiner also makes a gender observation:

> Women seem to have a finer sensitivity to their symptoms and to heed them more. Often men who attend our clinic will say 'I have *no* problem, but my wife really suffers.' Then we do some lab tests and discover both are being biologically affected equally.

A short wilderness retreat, if away from cell antennas, can give you useful information. Dr. Scheiner suggests removing yourself from some of the sources for a few days, and seeing what happens. If the headaches, insomnia, tingling and other symptoms improve when you're away, and return when you come back, this is probably a sign that you're electro-sensitive.

Cindy Sage of the BioInitiative Group, agrees:

> Symptoms quickly improve when away from EMF/EMR sources, particularly when the patient moves away from computers, fluorescent lighting, transformers, wireless antenna, cell and cordless phones, appliances and out of proximity to cell phone towers, electrical substations and power lines…
>
> Symptoms recur on returning to the irradiated environment. Over time, sensitivity is increased to smaller and smaller EMF/EMR exposures.

The renowned surgeon Dr. William Rea, offers the EHC-D protocol used at his Environmental Health Center – Dallas (www.ehcd.com). This facility is a complete testing and medical treatment clinic for environmentally-sensitive adults and children.

1. Eliminate exposure
2. Energy balancing
3. Mineral analysis
4. Skin test for minerals
5. IV nutrients twice weekly
6. Grounding at least 1 hour/day
7. Immune modulation
8. Sauna

Katharina Gustavs says, "Grounding and sauna, as mentioned above, are great therapeutic strategies for healing. However, if the grounding involves artificial grounding pads, the electric field exposure could increase during the grounding. Walking barefoot along the beach, of course, is always a winner. Infrared saunas are notorious for high magnetic and electric field exposures."

One company that has been recommended has evidence that its low EMR emission infrared saunas are specifically designed for electro- and chemically-sensitive people. (See the Resources section.)

Sarah Dacre has greatly reduced her disabling EHS symptoms – including severe short-term memory loss, with a variety of interventions:

> After a few years of dealing with this, I now have 95% of my perfect memory back. How did I achieve this? By shielding my life from EMR; moving my bed; switching off all EMF emitting devices and detoxifying. I drink only spring water and eat a very healthy home cooked diet, and take pure supplements and minerals.
>
> I sleep on a foam mattress supported by a wooden bed. I have regular Chinese acupuncture where the remedial focus is on memory and a healthy central nervous system.
>
> Additionally, I no longer use a mobile (cell) phone, which caused me withdrawal symptoms. I went through cold turkey from this addiction. My right eyesight has since improved enough not to wear glasses.

I walk in nature every day both for the exercise and fresh air, oxygen is vital to assist my body cope with its heavy toxic load.

The WEEP Initiative has a section on living with EHS on its website, www.weepinitiative.org/livingwithEHS.html

Now, let's keep going and look at a few more conditions and explore how EMR exposure influences them. You may discover some connections with health issues in your family, and some useful medical advice.

Our library mum – a woman who has been reading parts of the book as they've been written – has noted some interesting interconnections:

> It seems EMR is one of *various* environmental toxins that we are exposed to which are compromising our immune system.
>
> A couple of years back, I was exposing myself to a lot of bleach and when I stopped this, I recovered much of my health (slowly, over time).
>
> The challenge is that it is difficult to put our finger on *which* environmental toxin is the culprit because it is the overall load of toxins that get us...we need to abandon the search for a single cause.

Immune System Impairment

The Sweet Smell of...Chemical Overload?

Why should you care about this? We only have one immune system, one lymphatic system, one liver, one set of kidneys etc, to deal with the daily onslaught of environmental toxins, EMR now included.

Your immune system, whether you know or not, is reacting to all of these environmental exposures. The more we can reduce the various chemical loads, the better it is for us, and for all life forms on this planet, including the birds in the air, and the fish in the rivers and oceans. And most people who are electro-sensitive are also chemically sensitive, so this is vital information for them.

I asked Dr. William Rea what is of most concern to him these days. His answer: "Pesticide exposure plus wireless equipment. EMF sensitivity is a growing problem in the world. Everyone needs to be aware of its consequences."

Dr. Rea elucidates on the interplay of environmental influences:

> Various factors (heavy metals, inner terrain – yeasts, pH balance, emotional stress etc.) may combine to make us more vulnerable to a myriad of environmental toxins, including EMR. Dealing effectively with these factors can boost our ability to buffer the toxic effects. If the load is reduced, many times the patient can tolerate the EMR.

Systems Overwhelmed?

A recent report on the rapid rise of serious food allergies in children (including life-threatening ones like peanuts) tells us that the experts cannot explain the nearly 20% increase in the last decade. Could it be the deluge of chemicals in our air, food and environment are overwhelming young immune systems?

Some companies were recently defending the several carcinogens found in their baby lotions and bath products, as just 'trace amounts'.

I have also discovered one also needs to be vigilant for the wolf in sheep's clothing: chemically-laden products are labelled "natural" or "fresh", or some such disguise.

A story comes to mind: when I had a counseling practice, a woman told me that she had used a plug-in air freshener for her son's bedroom to purify the air to help with his asthma. Like most of us, she was getting her health information from marketers.

You have probably heard that lead, tar and formaldehyde are used in most cosmetics and 'personal care products'. When they say 'perfume', unless they are using only essential oils, assume these are chemicals, as in most scented candles, cleaning products, air fresheners. Taking these things out of your home can ease everyone's immune and respiratory systems.

Here's some more good information from Larry Gust:

> Fragrances commonly use phthalates, some of which are hormone disrupters. These, as well as synthetic musk compounds and other chemicals,

accumulate in human tissue and are found in breast milk. Manufacturers don't necessarily test this stuff for neurological effects or respiratory effects – even though we know an increasing number of people have asthma attacks when exposed to scented products and two fragrance materials that had been used for decades were found to be neurotoxic.

Can we live in a pure, non-toxic environment? I don't think so. And, yes, our immune systems need a good workout – so we shouldn't swab down the decks with anti-bacterial soaps.

However, given the levels of chemical and electric pollutants in our environment it seems prudent to reduce the exposures – change what we can – and not agonize too much about the ones we cannot. And boost our ability to handle these. Something like a healthy household serenity prayer?

Cleaning Up Air Pollution in Your Home

Home air is contaminated by organic and inorganic particles and toxic gases. Organic particles can be mold and mildew spores, dust mite parts and feces, animal dander, fabric dust and tobacco smoke.

Inorganic particles include very fine soluble and insoluble dusts generated by industrial processes which are carried everywhere by the wind and find their way into homes.

Toxic gases include leaking natural gas and volatile organic compounds (VOCs) produced by man-made and natural sources. Particle board, new furniture, carpet pad, and new permanent press clothing release formaldehyde, a potent allergy sensitizer and known carcinogen. People most at risk from indoor air pollution are newborns, young children, the elderly, heart patients and those with bronchitis, asthma, and allergies.

To eliminate as many pollutant sources as possible from your indoor environment consider the following helpful hints:

1. Vinyl flooring, shower curtains and water bed bladders release phthalate gases. (Phthalate oils are used to make these PVC products flexible.)

2. Chemical air fresheners including plug-ins are neurotoxins. Studies show exposed mice exhibit nerve damage and even death.

3. Mold and mildew (fungus) also produce detrimental VOCs as by-products of living. Therefore, heavy fungal

contamination in or below living areas is to be avoided.

4. Air out dry cleaning in the garage.

5. Read labels on cleaning products. If it says "Use in well ventilated room" it means that breathing the vapors can be unhealthy. Look for alternate products at your health food store.

6. If your garage is below an occupied area, leave the door open until the car engine cools off or leave the car out until it cools off. Get oil and gas leaks fixed.

7. Store gasoline cans and equipment with gas engines in a area separated from the house.

8. New permanent press clothes can be soaked in the washer with two cups of powdered milk for several hours then rinsed and washed normally.

9. During times of high humidity, indoor conditions become ripe for growth of fungi (mold, mildew) and dust mites.

   Suggestions to prevent fungal problems:

   a) Move furniture away from wall to allow air to circulate behind.

   b) Keep closet doors open to allow air circulation around the clothes.

   c) Make sure closet is uncluttered enough to allow air to circulate.

   d) If you have a problem, place a low wattage bulb (not fluorescent) in closet. Keep bulb away from clothing and always on. Most fungi don't like light.

e) Treat infected areas with one cup bleach to one gallon of water. Wear rubber gloves and eye protection. For dust mites wash bedding and vacuum the mattress and carpeting often.

10. Water leaking into the house from outside or plumbing leaks can have serious health consequences. Wet areas must be dried within 24 hours or rapid fungal growth creates masses of spores and hydrocarbon gases. Make sure your insurance company pays for a proper clean up.

11. Air filters that remove fine particles and/or gases are available as room filters, or as whole house filters that attach to the central heating/air conditioning.

    The best type of air filter is the true HEPA (high efficiency particulate arrestor) which removes 99.97% of particles over 0.3 millionths of an inch in diameter. If the filter is not a true HEPA it does not remove the finer and more harmful particles which makeup 92% of the dust in the air. Standard furnace filters are useless in protecting you from fine particles. When buying, check the price of replacement filters and the change interval against the original cost of the equipment.

12. Standard vacuum cleaners return 70% of the dirt sucked up right back into the air. Vacuums with a true HEPA filter are now made by several manufacturers.

    Dirty air stresses the body and takes energy away from rejuvenation. This is particularly true during sleep time when your body is repairing itself.

    Larry Gust, Gust Environmental, www.healbuildings.com

• IBE Certified Building Biologist • IAQA Certified Mold Remediator

Dr. Mark Hyman is a physician who emphasizes the benefits of addressing environmental pollutants, including EMR:

> Clearly there are many environmental influences that can have a major impact on overall health, including your brain health.
>
> These toxins affect us all from our food, our water and air, medications, chemicals, metals, molds and electromagnetic radiation.
>
> Everyday choices we make to protect ourselves will have impact globally. The European Union, through new legislation (called REACH) designed to limit toxins in commercially produced goods, caused China to change its manufacturing practices.
>
> Buying organic food, refusing vaccines with mercury, or buying deodorant without aluminum will create a ripple effect throughout our economy and environment.
>
> You can make changes in your life and in your world with the choices you make. The choice is yours.
>
> What's more, you can make changes in your personal health.
>
> You *can* heal yourself from brain damage by altering your diet, limiting your exposure to toxins, and changing the way you live. You can REVERSE the effects of depression, anxiety, bipolar disorder, autism, Alzheimer's, ADHD, and more if you know how to.
>
> From *The UltraMind Solution,* (Simon & Schuster 2009)

Chronic Fatigue Syndrome (CFS) In the UK, this condition is called myalgic encephalomyelitis (ME).

As you may recall, throughout this book, several of our experts have advised us that EMR exposure weakens the immune system. This is not news.

An overwhelmed immune system – our body's wondrous system of protecting our health – cannot fend off environmental toxins, bacteria and viruses leaving us more open to colds, flu etc.

Not only do we get sick more often, it can take us much longer to recover. And we feel more tired, more often.

A compromised immune system is also a contributing factor in, and a result of, Chronic Fatigue Syndrome, and many other conditions. (As you may know, emotional stress also weakens our immune systems.)

While CFS/ME is a complex illness with many possible causes, in October 2009 an American research team reported, in the journal *Science*, that a single retrovirus – a blood-borne pathogen – may play a major role. Retroviruses are known to cause neurological symptoms, cancer and immunological deficiencies, and many people with chronic fatigue report their symptoms emerged after a serious viral infection. While it's too early to say whether this discovery will lead to better treatments, the news is heartening to the millions of people worldwide who suffer from this debilitating condition.

The reason I am including a section on this is that CFS and exposure to electro-magnetic radiation share many symptoms and are connected. And knowing about CFS can be a great boost to your overall health and your ability to heal from, or ward off, the effects of EMR.

In 1983, W. Grundler, in Germany, found even very low levels of EMR exposure affects the increased growth of yeast cells.

("Sharp Resonances in Yeast Growth Prove Nonthermal Sensitivity to Microwaves").

Dr. Rau adds his recommendations for dealing with electro-sensitivity and fighting yeast conditions:

> Cultures have shown beneficial bacteria grow more slowly in the presence of electromagnetic fields allowing pathological (yeast – candida) organisms to dominate.
>
> Thus, a strategy with electrically sensitive patients, or with those facing chronic conditions, is the aggressive supplementation with probiotics and other Biological Medicine approaches to balance intestinal flora.
>
> Many people with chronic infections likely linked to EMR exposures, such as Lyme Disease, are symptom-free after an aggressive microorganism rebalancing program.

A proliferation of yeast cells (Candida albicans) in the bowel can be a contributing factor in many conditions, including Chronic Fatigue Syndrome, and will weaken the health of EHS people.

A physician specialising in complementary medicine, Carolyn Dean MD ND, author/coauthor of 17 books including *The Magnesium Miracle,* explained in a recent interview that it's not just fatigue:

In brief, Candida albicans is a fungus living in our intestines that produces 180 chemical toxins capable of making you feel dizzy and fatigued, shutting down your thyroid, throwing your hormones off balance, and causing you to crave sugar and alcohol, and gain weight. It's associated with drowsiness, insomnia, numbness, tingling, PMS, loss of libido, painful intercourse, infertility, MS, Crohn's, colitis, IBS, acne, Lupus, white tongue, bad breath, body odor, sinusitis, bruising, sore throat, bronchitis, shortness of breath, heart palpitations, spots in front of eyes.

The miracle of antibiotics has its downside as an underlying cause of yeast overgrowth. The refining of sugar and wheat has its downside by creating a simple food source for yeast. The tremendous levels of stress hormones that flood our bodies daily also make us prey to yeast. When yeast, bacterial, and food toxins hit the blood stream they trigger widespread inflammatory reactions by either directly attacking tissues or creating allergic reactions along with the production of histamine.

It is mainly through diet and lifestyle change that you can overcome yeast overgrowth—but you *can* overcome it, and reduce the amount of inflammation in your life with some effort and support.

Original article: Total Health and Longevity Magazine Sept 2006 issue.

Dr. Dean also describes how magnesium can be useful in CFS and possibly EHS as these conditions share many symptoms, including insomnia and fatigue, as we have seen:

Magnesium-deficient patients commonly experience fatigue because dozens of enzyme systems are under-functioning. An early symptom of magnesium deficiency is fatigue.

Dr. Dean is the medical director of the Nutritional Magnesium Association and has a wellness consulting practice in Hawaii.

The treatment suggested for CFS, and boosting one's immune system, includes cutting out sugar. Sadly, the more difficult this is, the more it may be needed, as craving sweets is one of the main symptoms.

Physicians specializing in this also recommend supplementing with good quality probiotics – it's not enough to eat yogurt – and using an effective anti-fungal. There are more details at Dr. Dean's website: www.drcarolyndean.com

Another physician specializing in complementary medicine, and an expert in CFS, Dr. Jacob Teitelbaum has more information on the treatment of CFS on his site www.endfatigue.com

I asked Dr. Teitelbaum this question: "Many people, including those with EMR exposure, are tired and don't sleep well. So how do you know if you really have CFS?"

> Although many people are tired and have poor sleep, the fatigue usually goes away with rest and is not associated with the other symptoms. It is the mix of a common group of symptoms that define a syndrome.
>
> People who are tired, achy, 'brain foggy', can't sleep, have increased thirst and poor libido – even if these are not disabling – often have a milder form of CFS.

Electro-sensitivity and CFS

Dr. Teitelbaum offers his responses to a patient experiencing sensitivity to electric fields and computers.

Q. "I had a relapse with CFS and have been out of work for a year. I notice that this time around a few months into the illness, that I am now sensitive to sitting in front of a computer screen. Symptoms: I feel hot and nauseated, and if I stay on the computer long enough, feel my heart racing. I notice that I'm really sensitive to it in the morning versus the evening. Any ideas?"

Dr T. "There are those who get electro-sensitivity with CFS. In some cases, turning off the electric circuit to your bedroom before you go to bed (use a wind up alarm clock or a battery one kept away from your head) and leaving the electricity off at night (you would be amazed at some of the electromagnetic fields found in the bedroom) can also help – especially since you report feeling worse in the morning.

In severe cases, there are specialists that can evaluate a home for electrical pollution and assist in remediation."

Sleep, perchance to dream?

The curtains are drawn tightly. The spacious room is dark and seemingly conducive to sleep. The woman in the brass bed has one arm twisted around the sheets; the other's flung out the side and rests lightly on her wireless laptop. On the night table the digital clock marks the minutes of another sleepless night.

Her trouble falling sleep, and then staying asleep, seems to be getting worse. "Stress, no doubt," the doctor advises and prescribes pharmaceuticals. Annoyingly, these seem to make her dozier during the day than they do at night. These pills perch on the table along with an arsenal of headache remedies. She read that room temperature is crucial, so, she either snuggles a heating pad, or relies on an air conditioner. Nothing is working. Ever resourceful, she has surrounded herself with the latest techno-delights. In the middle of the night she often she flicks on the light and taps away on her laptop, or picks up her cell phone and calls a friend in an earlier time zone. She likes its all-in-one convenience – she can shop online, plan vacations, stay in touch, and the alarm is so pleasing she tucks it under her pillow. There's also a cordless phone on her nightstand; being an early adopter she has the power-packed DECT 6.0 model, accessorized with a touch-screen for the ultimate in wireless chic. She no longer feels sleepy. Then, she looks over and sees the clock …

Insomnia, Stress, Mood

As you may know too well, not being able to sleep affects every facet of your life to the point where you can barely function. Specialists say it doesn't take long before sleep deprivation drives down your IQ (temporarily!), impairs reasoning and memory, makes adults more irritable and kids a little more "hyper".

As if that isn't enough, scientists also suspect that chronic sleep deprivation in children may also impair their growth and development. There are of course many factors that contribute to sleep disruption – caffeine, stress, hormones, a bedroom that faces a noisy street…

Not too long ago, our family was staying at a hotel when I attended a conference. I had chosen this hotel because it was one of the few that did not have wireless Internet access in the room.

My relief was short-lived, however, when I got into our room, looked out the window, and saw a nearby cell tower antenna beaming down on us. I had with me the small Sensory Perspective's 'Electrosmog Detector', which has a very sensitive audio signal. The EMR levels were so high that this EMR warning system screamed as loud as having a cell phone in use.

It wasn't easy explaining to the front desk why we wanted to change our room. I found a room on a lower floor, with less exposure, but when we left the hotel, I felt a pang of sadness for people seeking help in the sleep research centre that was on the top floor of this building. (Also I felt for the housekeeping staff with powerful vacuum cleaners strapped to their backs. Intensity. Proximity. Duration.)

In stress disorders and depression, EMR has been shown to affect levels of brain chemicals and hormones including norepinephrine, dopamine and serotonin, which are involved in regulating mood. As you may know, these conditions are already epidemic so more research is needed here. Maybe this why we feel more calm, and our spirits are lifted, when we are out in nature and far removed from technology.

As Dr. Scheiner describes, your ability to sleep can be a good indicator of your EMR exposure and sensitivity.

Many people who are not sleeping well may be unknowingly immersed in a high EMR environment. We know the agitating effect of EMR on the brain. Is it any wonder that difficulty sleeping is one of the primary and most prevalent symptoms? Evidently, some people are using electric mattress pads to sleep better, and/or to enhance their health. You may remember reading about why this is a problem, in Step 3.

EMR exposure has been shown to make it more difficult to "wind down" and can also reduce the amount of melatonin in the brain. According to some research, using a cell phone for more than 25 minutes a day for two weeks can be enough to reduce melatonin levels. Katharina Gustavs adds, "According to J. Burch's research, the effect is exacerbated if you sit in a dimly lit office for most of the day."

As you may know, this hormone is necessary for deep, restful sleep, which also allows the body to recharge and heal itself. (Many years ago, a remarkable immunologist told me that one of the primary functions of sleep was to recharge the immune system.)

A small study done by Sweden's Karolinska Institute and Uppsala University, with Wayne State University in Detroit, released in January 2008, revealed that people using a cellphone just before bedtime took longer to get into deep sleep, and spent less time in this deeper sleep phase.

(Published by the Massachusetts Institute of Technology's Progress in Electromagnetics Research Symposium).

One of the researchers, Dr. Bengt Arnetz, said it appears that cellphones affect the areas of the brain responsible for activating and co-ordinating the stress system. It's also possible that radio waves disrupt production of melatonin, which controls the body's internal circadian rhythms.

Melatonin performs many crucial functions including regulating the timing of the onset of puberty.

Therefore it's no wonder sleep experts such as Dr. Jeffrey Lipsitz were especially disturbed by what this study implies for teenagers, who tend to use cellphones more in the evening, talk for long periods of time, and keep the phones on standby during the night.

As he told CTV News, a Canadian TV network:

> If they're using cell phones for long hours in the evening and then going to sleep and their sleep is disturbed this may have implications with regard to health and development and functioning in school and so on...

Insufficient sleep in young people is linked to mood and personality changes, ADHD-like symptoms, depression, lack of concentration and poor academic performance.

Dr. Lipsitz went on to say:

> You kind of wonder what else might be going on to the brain as a result of extended cell phone use, and what does that mean for all of us?... It certainly cries out for more research.

(Incidentally, the study was funded by the Mobile Manufacturers Forum, which called the results "inconclusive.")

Dr. Jacob Teitelbaum also writes about insomnia in *From Fatigued to Fantastic* and recommends: "magnesium 75 to 250 mg and calcium, 600 mg at bedtime and Hydroxy L-tryptophan (5-HTP) 200 to 400 mg at night. 5-HTP is what your body uses to make serotonin, a neurotransmitter that helps improve the quality of sleep." It may take four to six weeks to see the full effect of 5-HTP, though, so don't expect instant results.

Dr. Teitelbaum cautions that if you're taking anything else that raises your serotonin levels (including SSRI-type anti-depressants or even herbal remedies like St. John's Wort), "it is reasonable to limit the 5-HTP to 200 mg at night." In rare cases people may experience a reaction from excess serotonin.

As with all supplements that affect the body's natural production of hormones, for example, you should consult with a qualified clinician – physician or naturopathic doctor – before following any protocols. There may be contra-indications, conflicts with certain conditions and/or medications you need to know about.

Dr. Teitelbaum also has some useful insight:

> For most people with disordered sleep, supplementing melatonin to normal levels takes 3/10^{ths} of a milligram. The usual dose is 3 milligrams, which is ten times the level I recommend.

I have found that sleeping in a dark room, and one that stays dark until you want to wake up, is very helpful. Try blackout curtains, or good eyeshades. Did you know that eyeshades keep the light out of your eyes and help in sleeping, but you need the whole room darkened to enhance the production of sleep-inducing melatonin? And, of course, make your bedroom a low EMR sanctuary.

Sleeping Well Beyond The Electronic Cocoon

This once-sleepless mom heard our wireless wake-up call – loud and clear – and took action right away:

> I am a recovering breast cancer patient, and a life long skeptic, but after reading *Radiation Rescue*, I found the information in the book so compelling, that I decided to make some immediate changes.
>
> To start, I replaced my cordless phones with land lines, moved my blackberry charger out of my bedroom, moved my bedside clock across the room, and stopped sleeping with the blackberry on my night table. We removed all of my daughter's electronics from her room, and got rid of her metal bed frame. As a family, we decided to try out a self-imposed one-hour-per-day-limit on how much time we would spend

in front of the computer, television, DS and Wii for a month, and see what happened. We recognized that there may have to be adjustments to computer time made to manage work and/or school obligations. To be honest, while it seemed like the right thing to do, I didn't expect to notice any immediate effect. However, I have been struggling with insomnia for the past three years, two years prior to my breast cancer diagnosis. The first night after clearing all of the electronics out of my room, I slept like a baby. Likewise for the next night. While I am willing to consider a certain amount of placebo effect, I have now been sleeping well and deeply for a few weeks. I notice that I rarely wake in the night to use the washroom (before, I would get up several times in the night) and when I do wake up, I fall easily back to sleep.

My daughter, who at 13 has been complaining about not sleeping well for the last year, is also sleeping more deeply. Now that I think about it her sleeping difficulties began about the same time we put a cordless phone in her room, and mine began when I got my blackberry. Maybe just two coincidences. Or maybe not.

During my period of sleeplessness, I stopped drinking caffeine, and did everything I could to improve the quality of my sleep. I looked at all of the possible causes - new job, peri-menopause, caffeine intolerance, food allergies. I worked with a naturopath. While there was some small improvement, nothing was as dramatic as the effect of lowering the EMR levels in my bedroom and house.

My son has ADD and learning disabilities, and we wonder what other benefits our experiment in "unplugging" will reveal. Last night, we went to the park and played tennis after school, then had a rousing game of UNO after dinner. Playing cards and tennis are much preferable to the poor habits we had been slipping into, with each of us isolated in our individual electronic cocoons.

Cardiovascular – heart disease, TIAs, stroke

Cardiovascular disease is a major problem in industrialized countries, and it's not just because of the unhealthy western diet. Researchers explain that when EMR depletes the levels of antioxidants in the blood, high-density lipoproteins (HDL), the good cholesterol, will bind with free radicals (oxidants) turning the good cholesterol into bad cholesterol (LDL). This leads to thicker, "stickier" blood, which can form clots. (Emotional stress also thickens the blood, and constricts the arteries.)

And EMR has been shown to disturb our own sensitive electrical system – the brain and the heart, for example. I have a friend who is a scientist at a leading university. Despite being fit, and in otherwise excellent health, she was rushed to emergency in cardiac distress. There's a claim by staff that a cluster of cancers is related to the high levels of EMR on her campus.

Could the cardiac symptoms also be related? Some electro-sensitive people can actually tell when there is a cell phone powered on nearby – they feel a change in their heart rate and rhythm, and/or a tightness in their chest.

Years ago I co-produced a program for Canadian network TV on heart health. I'd like to reconnect with the hundreds of patients I counseled when I worked in cardiac clinics, and encourage them to read this research and do an EMR audit of their homes and workplaces. And I'd like to go back to all the cardiologists I once knew to ask if they're seeing an increase in "unexplained" cardiac symptoms.

In November 2009, Dr. Havas presented research demonstrating increased heart rate from wireless radiation exposure. Evidence included an increase in heart rate on exposure to a nearby cordless phone. The heart rate immediately returned to the baseline after that phone was unplugged (see www.magdahavas.com).

There is a concern that many patients today who are experiencing cardiac, and other electro-hypersensitivity, symptoms are turning to prescription drugs, instead of reducing their EMR exposure.

I welcomed this pioneering study and the opportunity to ask cardiologist Dr. Stephen Sinatra for his insight. He told me that he applauded Dr. Havas's research and offered this observation:

> We saw these cardiac effects ourselves in the middle of a study when a cordless phone was in use. If electro-pollution stresses the body like emotional stress then we already have mountains of evidence on how this affects our health including, of course, the cardiovascular system.

Ophthalmological/vision problems

EMR has been shown to damage the lens of the eye. According to Powerwatch, this has been known for more than forty years:

> In 1964 Dr. Milton Zaret was one of the first scientists to speak out about the dangers of microwave radiation and ocular effects. According to Zaret, exposure to this radiation at either thermal, or non-thermal, levels can cause cataracts which sometimes remain latent for months or years. Zaret then had his laboratory research funds cut off.

Metal eyeglasses can attract EMR, magnifying its effects. (Here's another tip from an eye surgeon: blue-tinted lenses are not recommended as they let in harmful rays of the light spectrum.) Another expert has told me that our eyes are one of the tissues that are especially endangered by exposure to EMR. The other particularly vulnerable body parts are the testicles.

Infertility, miscarriages, birth defects

For couples longing to conceive, all this is deeply distressing. Some researchers in human reproduction are now telling us the more time men spend on cell phones, or carry powered-on phones in the pockets of their pants, the lower their sperm quality.

Other work exposure to EMR may also be a factor in men's reduced fertility. A study of 17 thousand military employees in the Norwegian navy, between 1950 and 2002, found men who worked within ten miles of high frequency aerials had a significantly higher risk of infertility than the ones who worked farther away. (Baste V, Riise T, Moen BE Eur J Epidemiol 2008)

In Japan, with its massive concentration of electro-smog, an unusually high number of women are suffering miscarriages in their last trimester. (Nagaishi et. al., J.*Obstet Gynaecol Res.* 2004)

In August 2009, Dr. Louis Slesin of Microwave news reported:

> Martine Hours, the chief science advisor to the French RF research program, has called for serious studies on the effects of cell phones on male fertility. Now, Australia's John Aitken is saying the issue 'deserves our immediate attention.' In a new study, Aitken has reported that cell phone radiation damages human sperm – as well as DNA. Cell phone use and radiation has now been shown to harm sperm, in five different countries and including two different groups in the United States, one of which is at the Cleveland Clinic. Aitken's message is simple: Men who want to have children should not keep active mobile phones in their trouser pockets.

When couples trying to conceive know that EMR exposure decreases fertility they can now make informed decisions.

More girls than boys in your play group?

There are also indications of a global prevalence of female births. Some researchers suspect this may be due to EMR. A Russian study in the 1950's found an altered sex ratio (more girls) in children born during the years of a radar station's operation in what was known as Latvia. That Norwegian navy study, which I mentioned above, showed similar results: a lower ratio of boys to girls were born to fathers who reported a higher degree of exposure to high-frequency antennas and communication equipment.

 Evidently, the normal sex ratio is commonly 105 boys to 100 girls. But under adverse conditions – famine, for example - more girls are born. While this girl/boy imbalance may not seem troublesome, think of the ramifications over a few generations. And, perhaps we need to connect the dots with the growing signs of dysfunction throughout our ecosystem?

As I see it, information is the key to waking up, to getting ourselves out of this biological experiment, as Dr. Salford and Professor Johansson have cautioned us.

Neurological issues

Alzheimer's, dementia, Parkinson's Disease, ALS and MS may also be related to EMR, according to neuroscientists who explain this exposure can cause levels of neurotransmitters such as acetylcholine to drop abnormally. Low levels of acetylcholine are linked to a number of these neurological and neuromuscular symptoms and disorders.

(A side note: many people with these conditions improve dramatically when they deal with heavy metal toxicity and avoid artificial sweeteners such as those found in diet pop. Research the health effects of Aspartame, for example.)

You may not want to hear this; I know I didn't. While Alzheimer's Disease and dementia have generally been diseases of the elderly, based on the evidence of EMR biological harm, some experts are predicting these conditions will be afflicting people decades earlier.

The silent epidemic?

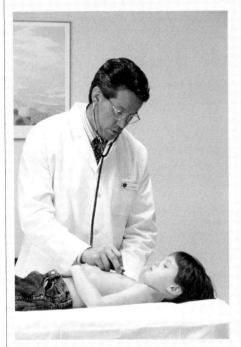

The young boy lies on the examining table banging his legs in an odd rhythm. He seems off in his own world. Standing close by, his parents look stunned: the pediatrician has just told them their child has autism.

They have known for a long time that something was not right but nothing prepared them for the raw shock of the diagnosis they had feared.

He had been such a healthy baby, and a normally developing toddler. Not long after his second birthday, however, they began to notice it was difficult to get him to focus, the development of his speech slowed, and he had occasional intense tantrums and hyperactivity.

They have been taking extra steps to protect him, including having a wireless baby monitor on at all times. They keep his environment as clean and germ-free as they can with highly effective cleaning products.

They are not giving up on their son; the mother leaps into action and learns everything she can. Her research leads her to an autism support group that she and her husband attend.

She reads about a link between autism and the mercury in childhood vaccines, and returns to quiz the doctor about the round of vaccines her son had around his second birthday.

The doctor gives her that dismissive look she will come to recognize all too well, and discounts this theory explaining autism has increased dramatically in California, where mercury has been removed from childhood vaccines.

Then she discovers another possible link – the increased levels of electro-magnetic radiation from cell towers and other exposures. They live in a very high-tech wireless-enabled community, and both she and her husband work in the computer industry. This makes more sense when she hears some encouraging news: children like her son have done well in a low EMR environment. And then there's a new study showing promising results.

ADD/ADHD – and the Autism Spectrum Disorder

If you are facing these challenges you don't need anyone to tell you how difficult this can be, especially in the initial stages.

However, many families, through their courage and determination to find out as much as they can, are improving other parents' success in helping their children. Truly dedicated parents who are leading support and advocacy networks are helping a great many families.

This section includes the research and recommendations of several leading experts in this field.

Here's what I've discovered: what most people call autism is a wide and varied range of neurological conditions with symptoms that may include difficulty communicating and functioning socially, various developmental disabilities, and cognitive challenges. Some people on the Autism Spectrum also have extraordinary musical or mathematical skills.

Not long ago I met a remarkable young boy who introduced himself right away, announcing, "My name is David and I have OCD (obsessive-compulsive disorder) and Asperger's."

His eyes were bright and alert; he was obviously highly intelligent, and he seemed agitated. David then became distressed when someone asked me about the book I was writing on the health effects of electro-magnetic radiation.

"Radiation?" he squealed, "Oh no. I don't want to know about that. That's too scary."

Many well-controlled adults have had the same reaction, but this vulnerable boy was visibly distraught. I shifted gears immediately and sat down with him and his mother, to listen to his fears, and do what I could to calm him. He appeared to feel better when we talked about all the ways we can be safer.

Evidently, David's parents had recently installed a DECT 6.0 cordless phone in his bedroom; he told me that he loves to talk for hours with his friends. He is a talented musician who spends many hours on an electronic keyboard, as well as a wireless laptop. Being environmentally concerned, the parents had switched all the lights in the house to compact fluorescents.

David's mother had just got the new high-powered 'smart' cell phone, with wireless headset, and told me that she had it on most of the day, and that it was often charged on standby in their bedroom at night – "just in case there's an important call". She had been considering getting her son a cell phone as "it would be helpful to know where he was, and so he could call us if he was in trouble".

Having read this far in the book, are you also concerned about this child, and many like him?

It seems a great deal of research still needs to be done into the ways EMR exposure, and heavy metal toxicity, may affect normal neurological development in children like David.

What are some researchers and clinicians telling us now?

There seems to be no medical consensus on what causes autism and other seemingly brain-based disorders like ADHD, or how best to treat them. Many researchers say there are probably genetic factors that predispose a child to such conditions. But some are now piecing together a disturbing picture of how environmental toxins, heavy metals, molds, chemical/food sensitivities, infections, viruses, and EMR exposure can combine to create symptoms.

As you've seen in the symptoms chart a few pages back, concentration problems and hyperactivity are among the symptoms of electro-sensitivity, so it may well be that many thousands of children are being diagnosed and treated with drugs when removing them from EMR exposure could help them significantly. Even children who don't have neurological problems can find it difficult to sit down, focus and succeed at school when there is a lot of EMR in the environment. (So why are so many schools installing wireless Internet?)

Some of the research that's coming in suggests a link: an American and Danish team of university researchers studied 13,000 young mothers who used cell phones during pregnancy. Their results, published in an epidemiology journal in 2008, found these women's children had a 50% higher than usual incidence of hyperactive behavior and emotional problems.

Having read this far, you are probably now aware that the radiation from the mothers' cell phones is likely only one factor.

A hospital in San Diego, California, found that 14 out of a group of 18 autistic children studied had lesions in the brain identical to those in rats exposed to low electro-magnetic frequencies between one and six days after birth.

The medical pioneer, Dietrich Klinghardt MD, PhD, is a physician who believes that the increase in autism is connected to the proliferation of EMR and our increasingly toxic environment.

In his ground-breaking work, Dr. Klinghardt is finding out how to recognize and treat the effects of toxic EMR on the genes of a developing fetus. The hope is to prevent neurological and other problems from developing in the child.

In a recent interview, Dr. Klinghardt told me:

> In 10 of our autistic children we measured the EMR exposure of the mother in the location where she was sleeping before giving birth to the child – who later on would be diagnosed as autistic. We measured body voltage and microwave exposure from ambient background cell phone radiation.
>
> The results clearly showed significantly increased levels of EMF exposure in the affected mothers versus the control group.

Dr. Klinghardt has developed an approach to assessing exposure:

> We also take EMF/EMR measurements at the home (and whenever possible work location or school of children with ADHD, behavioral problems, developmental delay and in adults with severe neurologic disease (Parkinson's, ALS, etc.) and the

results are similar: the higher the exposure at home and at work, the more therapy-resistant or severe the illness.

The promising news is that many children are showing significant improvement in their symptoms, thanks to treatments that include: detoxification; supplementation with vitamins, antioxidants and cellular nutrients; dealing with food allergies and chemical sensitivities; infrared saunas, and a low EMF/EMR environment.

Katharina Gustavs adds, "As long as the infrared sauna being used is not a magnetic field hot spot."

Dr. Klinghardt's suggestions for reducing EMR exposure are not difficult and are affordable for most people. Here is the short list that is given to his patients, including those with ASD (autism):

a. no wireless technology and no cordless phones in the home (get wired broadband and corded phones)

b. switch off all fuses at night (unless you can measure body voltage and get specific)

c. use the sleep sanctuary principle: a silver coated mosquito net over the bed as a Faraday cage (greatly decreases microwave exposure)

Gustavs comments on using shielding fabric, "First, the electric fields in the room have to be measured – and reduced – to make sure they are not drawn into the Faraday cage increasing the radiation exposure."

Chris Anderson adds, "Please do the research and be cautious of what RF shielding fabric you purchase. There are cheaper varieties that tend to become brittle and shed off tiny particles of the metals used in the fabric, which can find their way into food and bodies and cause harm."

Generation Rescue, an international movement of scientists and physicians researching the causes and treatments for autism, ADHD and chronic illness, is aware of this issue:

> Our community is very concerned about the effects of Electromagnetic Radiation (EMR) on the body and how it deregulates the ability to manage toxins and infections, and how it seems to interfere with blood flow in the brain. We feel that Dr. Crofton's approach to reducing EMR exposure should be looked at by all families looking to raise a healthy child.

While evidence of the link between EMR and autism is still evolving, the experts I consulted praise the benefits of *all* children spending less electronic time, and more time discovering the simple wonders of nature. Is your son or daughter mostly an 'inside', or 'outside', child?

Children With Autism Symptoms Benefit From A Low EMR Environment

This interesting and hopeful case study comes from Larry Gust, the electrical engineer and certified IBE environmental health consultant based in California:

> In mid 2008 I was called by a client with two children who were diagnosed with high heavy metals concentrations – particularly mercury. There was a 5-year-old boy with autism, and his sister, 3 years old, was behaviorally normal.
>
> Under medical supervision various strategies were tried to remove this toxic load from the children's bodies, but the results (monitored via urine analysis) were poor. A full environmental assessment of the house showed the principal issues were electromagnetic pollution from:
>
> - A wireless Internet connection.
> - A 5.8 GigaHertz cordless phone system.
> - AC Magnetic field level of more than 3 milliGauss in the girl's bedroom.
> - AC electric field of more than 7 Volts/metre in all bedrooms.
>
> **These changes were recommended:**
>
> - Eliminate the wireless Internet and cordless phone system and replace these with hard-wired Internet connections, and corded landline phones.

- Locate and repair wiring errors causing high magnetic field.

- At night, turn off the circuit breakers identified as causing the high electric field level in all the bedrooms.

> Just before these changes were implemented, the children had a urine test to determine if heavy metal were being eliminated from the body to the urine. There was little evidence of mercury dumping.

> One month after the changes another test was conducted. Mercury dumping doubled and uranium was showing up in the urine as well. By late fall 2008 the son was doing better than he had ever done in all aspects including behavior and his ability to learn.

Perhaps this is the picture: lower EMR = increase in heavy metal release = reduction in autism symptoms.

This is what's called anecdotal evidence, as you may know. Scientifically, this is not considered significant; it is, however, if it is your child who is getting better!

Connecting the dots again, this is interesting in light of Dr. Carlo's explanation that EMR compresses the cell membrane affecting the transport channels, and inhibiting the release of toxins, including heavy metals.

EMR, Mercury and Autism?

You will notice that words like detoxification and mercury come up in this discussion on autism. This may come as a surprise to many researchers.

The still-forming picture of this mechanism is information that may benefit everyone who has issues with heavy metal toxicity, and may shed a new light on autism and other neurological disorders.

In his book, *From Fatigued to Fantastic,* Dr. Teitelbaum describes how the cell membrane works:

> All cells in the body are enclosed in a cell membrane, a balloon-like wall... Stiff cell walls can make it hard for cell membranes to perform their critical functions.

As I outlined earlier, researchers tell us when cells are exposed to EMR they go into a protective mode to ward off this foreign invader. Part of the reaction is compression of the cell membrane, making it rigid; this affects the in-flow of nutrients and out flow of toxins and waste products. This makes it difficult for cells to clear the mercury and other heavy metals.

Remember our river of health. With increased EMR exposure, the cells are not able to clear toxins properly, as we continue to dump more into the river, including repeated doses of mercury.

Boyd Haley, PhD, is the former chairman of the Department of Chemistry at the University of Kentucky and a National Institutes of Health post-doctoral scholar at Yale University Medical School.

Dr. Haley's research into the biochemical abnormalities found in people with Alzheimer's disease led to his identifying mercury toxicity as a major factor. He was also one of the first to suggest that Thimerosal, the mercury-based preservative in infant vaccines, was the most likely toxic substance involved in autism-related disorders.

As we know there are benefits of childhood vaccination. It is hard to understand, however, why any of these vaccines still contain the highly toxic mercury. "They are safe levels," we're told. (We're getting used to hearing about so-called safe levels.)

In a 2006 interview on Autism One Radio, Dr. Haley speaks about the connection between mercury, other heavy metal toxicity and ASD, "The data is getting overwhelming, and I don't know how long our government can ignore it."

Jenny McCarthy and Generation Rescue are advocating Green Vaccines – removing all the toxins including mercury and aluminum. There's a lot of useful information for families at: www.generationrescue.org

In addition to receiving vaccines, a child can also be exposed to mercury prenatally, through the mother's mercury amalgam dental fillings, or later on if the child gets amalgam fillings.

Heavy metal exposure can also come from the environment, as in the case of eating fish, particularly tuna, from contaminated waters.

According to Dr. Klinghardt and others who treat ASD, it appears that people with a genetic predisposition to autism have trouble clearing heavy metals from their cells.

An explanation for this, offered by some scientists, is that when a child's blood-brain barrier is more fully formed – around the age of 18 months – it can make it extremely difficult for the mercury from the vaccines received up until that date, and presumably any other heavy metal overload, to be excreted from the brain cells.

Dr. Jerry Kartzinel, MD, is a pediatrician who co-wrote, with Jenny McCarthy, the information-packed book, *Healing and Preventing Autism* (see Books in Resources, under Jenny McCarthy). Dr. Kartzinel explains why some children seem to be able to detox themselves from metal exposures – and not develop symptoms of autism – while others can't:

We have natural pathways that are designed to clear toxic metals from our body. We are all born with different capacities to accomplish this. Some of our population is more easily poisoned than others. We know the very young and the very old are especially sensitive to poisonings and heavy metal... if they were never exposed to toxins in the womb, they might never have had a problem. If they were exposed to just "regular" toxins of day-to-day living, they would be fine.

Remember, we are speaking of a total body burden of heavy metals and toxins. This burden starts just after conception. By the time these babies take their first breath, their ability to mitigate the damage of toxic exposures is already taxed. Some children are barely taxed. Some children are almost maximally taxed. Those children born with a robust ability to detoxify sail right through childhood immunizations and everything else this world throws at them. For others, though, clinical signs and symptoms start becoming evident at a very young age.

So when we hear that most kids get these vaccines and they're fine, my answer to that is that they were blessed with a detoxification pathway that works.

Even if your family is not dealing with any challenge of hyperactivity/autism, don't you find the increasing incidence disturbing? I do.

Even if the human side of this epidemic does not unsettle you, you can imagine the millions of dollars spent on assessment, treatment and support services.

Wouldn't it make sense to explore all avenues to seek the cause and cure of these conditions?

It seems to me that children with these neurological symptoms/conditions should be at the forefront of our efforts to reduce the EMR and other toxic exposures in our homes, schools, health care facilities and communities.

Detoxification

In *The Magnesium Miracle*, Dr. Carolyn Dean describes the many benefits of magnesium, including the process of detoxification and the ability of heavy metals to be released through the cell membrane:

> Magnesium is crucial for the removal of toxic substances and heavy metals such as aluminum and lead from the body.

This is interesting in terms of the treatment of autism; I have also heard that magnesium is being used by some clinicians for EHS patients as it 'softens the cell membrane'.

Dr. Scheiner agrees detoxification is a key element in helping the body withstand even quite severe environmental assaults. He points out that while people in Finland were heavily exposed to radiation from the nuclear meltdown in Chernobyl, they had far lower levels of radiation sickness, and cancer, than you would expect. That's because, he says, the Finns' lifestyle includes saunas.

Sweating is one of the body's best mechanisms for clearing toxins. These specialists recommend the far infrared sauna – it seems more effective at detoxifying and you don't get as overheated as with a regular sauna. As you'll have figured out by now, you want to sit away from the electric element and find a lower EMR model. See the Resources Section.

Dr. Teitelbaum also offers a key to detoxification and/or buffering the effects of EMR with advice on the normal functioning of the cell membrane:

This cell membrane is made up of fatty acids and a phosphorus molecule. Your body likes to use Omega-3 and -6 fatty acids when it makes cell membranes. When these are not available, your body has to use saturated fats and this results in rigid, poorly-functioning cellular walls.

Essential fatty acids (EFAs) Omega-3 and -6 fatty acids play an important part here in keeping the cell membrane optimally functioning. This is particularly important when people are under severe physical stress or have increased sensitivities, including those sensitive to EMR. It is recommended to supplement even a healthy diet with good quality Omega-3 and -6 fatty acids."

A comprehensive protocol is evolving here. Check our website for more details as they are available from the leading clinicians and researchers in this field.

While the comprehensive protocol combining the recommendations from all of our contributing experts is still a work in progress, there are many useful tips, including the benefits of magnesium from Dr. Dean, the essential fatty acids from Dr. Teitelbaum, the benefits of antioxidants from Dr. Scheiner and the therapeutic role of supervised detoxification from Dr. Klinghardt.

Nutritional Supplementation

Not all supplements are created equal, as you may know. Many are not absorbed by the cells of the body and quite literally go right through you. And with what we have learned about how EMR compresses the cell membrane, it is an even greater challenge for nutrients to be absorbed. I have heard some are working on supplements that can soften the cell membrane, including some forms of magnesium.

Dr. Scheiner also suggests several ways to strengthen and protect our immune system and overall health with supplements:

> First, avoid exposure of environmental pollutants, as much as possible. This will help get rid of excess free radicals. To combat these we need scavengers – antioxidant vitamins. It is not possible to get all the nutrients we need from even the healthiest diet.
>
> Our nourishment is losing antioxidants because of industrial agriculture. These days a strawberry has only 20% of its original amount of vitamin C. Therefore, it is necessary to use supplements including antioxidants, vitamins D and E, as well.
>
> All these water soluble vitamins don't enter the brain so we need scavengers that can pass the blood-brain barrier – melatonin, for example.

Other Key Health Factors

We've heard the experts talk about all the factors working together in a negative way, harming our health. But we can also combine many factors to contribute positively to our family's wellbeing, by boosting our bodies' abilities to buffer the electro-magnetic load we experience every day.

After you have reduced your EMR, and some chemical, exposures, here are effective ways to boost your overall wellness and ability to deal with the EMR load on your system that you cannot control.

Some physicians offer recommendations that can enhance your family action plan; they have become central to ours. We begin with dental health.

Dental Health

Dr. Klinghardt is only one of the physicians testing patients for heavy metal toxicity as one of the underlying causes of electro-sensitivity, Chronic Fatigue Syndrome, autism and many other conditions. Research the work of Dr. Boyd Haley, as well. This is the focus of many complementary medicine clinics, including that of Dr. Thomas Rau, the Swiss physician.

Dr. Rau advises:

> A strategy to consider for those experiencing electrical sensitivity symptoms is to remove the electromagnetic 'hot spot' in the head created by the presence of metal fillings.

Concern is thus not only for the neurotoxic aspect of mercury in fillings, an increasingly understood hazard, but because fillings themselves act as antennas in the presence of electromagnetic fields from cell phones and cell towers, Wi-Fi networks, portable phones, and other sources of radio frequency radiation.

Dr. Rau says the removal of dental fillings can be an important early step in reducing electrical sensitivity, allowing some people to live in homes they otherwise could not tolerate.

(NOTE: removal of mercury fillings must be done by a dentist using a strict protocol, otherwise you, and the dentist, will be exposed to the mercury during the removal process. You must receive supplemental air during the procedure, I am told.)

Dr. Myron Wentz, a medical research scientist with a doctorate in microbiology and immunology, is also an internationally recognized virologist and cell biologist. Dr. Wentz is the founder of the comprehensive medical clinic, Sanoviv.

Dr. Wentz has written a book detailing his concerns about mercury. The following is an excerpt from *A Mouth Full of Poison*, (Medicis, S.C. 2004):

> Today we know that mercury is the most toxic, non-radioactive heavy metal on the planet. Despite its toxicity, mercury is still used in manufacturing including the production of: batteries, thermometers, barometers, semi-conductors, electronic instruments, lighting and chemicals.

In addition, it is used in several industrial applications such as electrical and power production, mining, electro-plating and jewelry making. Mercury is also found in fungicides employed by the agricultural industry, and in antiseptic agents and vaccine preservatives used in modern medicine.

Mercury exposure, even at very low levels, diminishes the ability of the immune system to meet the slightest challenge. The number of T cells in the blood declines markedly when mercury amalgam is placed in the teeth, and the cells have now been programmed to undergo apoptosis – cellular death – in the presence of this mercury.

If you have personally decided to get vaccinated, or are required to do so, I urge you to opt for the nasal mist rather than the injectable vaccine. The nasal mist is free of mercury-containing thimerosal and is the natural route for administering a vaccine for a respiratory virus.

Biological Dentistry

John Snively, DDS, practised 25 years as a Biological Dentist incorporating complementary modalities of homeopathy, detoxification, supplementation and energetic medicine. Dr. Snively explains this approach:

> There is no area of our body that influences our overall health and wellness more than the dental component. Food digestion begins here as does all manner of communication. In the zone of the teeth

and jaws there lies a sophisticated network of blood vessels, nerves, lymphatic drainage and muscles.

All the major acupuncture meridians originate with the teeth and imbalances here either in occlusion (bite) or toxic fillings can have far reaching effects. Each dental procedure is a traumatic surgical event and leaves its imprint on our autonomic nervous system.

Toxic metals and other materials often become implanted in our teeth resulting in a long term burden to our immune system and root canals can also have devastating effects. The teeth and jaw joint provide us with a valuable sense of proprioception and discrepancies here can manifest throughout the body.

Dissimilar metals in the mouth can create not only an electrical current but can also act as an antenna for other invisible Electromagnetic and Microwave forces.

In seeking a Biological Dentist one must first be certain that the office is "mercury free" and provides protection not only for you as a patient but that the dental team is wearing appropriate masks when removing toxic metals.

A very thorough whole health/medical interview should be conducted to evaluate previous treatments, sensitivities and events including emotional.

Attention to diet, nutrition, detoxification procedures and lymphatic drainage along with herbal and homeopathic support should be available.

Some sort of biocompatibility testing should be offered to determine which materials are most tolerable for your immune system. Clifford Consulting & Research, of Colorado Springs, offers this testing – a dental materials screening (www.ccrlab.com).

A neighbor tells us about her journey with this wake-up call.

I am encouraged to see the awareness around cell phone use growing – following the Larry King Live program, and the columns in the NY Times soliciting reader responses. It seems users have a very limited, and often only intuitive, sense that there may be potential health concerns from the use of cell phones and other wireless devices.

As I am becoming increasingly aware myself, I am taking small steps to reduce my exposure and protect myself. I also am engaging friends and family, and even new acquaintances, around this topic. It concerns me, even pains me, to think about the harm we may be doing to ourselves. People need to know this.

Even still, I sit in front of my laptop for hours, doing Internet searches or email communication, with the wireless router on most of the time. I am becoming increasingly sensitive to bodily sensations that accompany, or follow, these stints.

Am I linking sensations and symptoms as I develop awareness of the health concerns – where previously I was simply ignorant? Or, maybe there is no link? I will be tracking this further. I don't want to be a hypochondriac, nor do I want to be wilfully naive and ignorant.

Safeguarding Your Family's Health the Holistic Way

As you may know, the complementary medicine approach has a lot to offer those of us who want to protect our families' health in this increasingly irradiated world. Complementary health practitioners tend to focus more on prevention and natural remedies such as those advocated by Dr. Andrew Weil and the others mentioned in this book, whereas the conventional medical model uses more pharmaceuticals to treat disease.

In this emerging field of electro-sensitivity it seems crucial to understand the complementary approach, as there are no drugs to treat this condition, and many conventional tests and treatments may exacerbate symptoms and conditions rather than improve them. Dr. Scheiner describes the electro-sensitive person who's experiencing a range of symptoms when exposed to these frequencies. Unfortunately, this percentage of the population is growing dramatically in virtually every corner of the world. Few people, and few of their doctors, recognize these symptoms and their potential cause.

Therefore, after many medical or non-medical tests and therapies, patients' symptoms may remain and over the long term even worsen, as they have not been able to identify and treat the source. For some, the source is the invisible waves of electro-magnetic radiation that surrounds us all.

The best possibility here is that people have recognized their body's wireless wake-up call, are hearing the body's alarm, and taking steps to avoid the exposure. While many people are suffering, with their daily activities severely limited, feeling overwhelmed and victimized by the electro-pollution, in some ways these canaries in the coal mine may be doing better than the miners, who are also being exposed and affected, but do not hear and/or heed the wake-up call symptoms.

One of the main objectives of this book, and my new health advocacy, is to get the word out that if you're having a lot of trouble sleeping, if you're fatigued, having skin tingling, burning sensations, difficulty focusing, any light-headedness or dizziness, this collection of symptoms may well be your body's attempt to let you know that this exposure is harmful.

Dr. Scheiner's observation:

> Each symptom, like the headaches, sleep disruption, dizziness etc. can be unspecific – caused by a lot of different reasons. It gets specific, however, when you experience several of these – this could be what we're calling microwave syndrome – this is not a sickness, it is a warning system.

Your body is trying to alert you, saying, 'I'm having trouble here, get me out of this situation'. These symptoms are a call for help.

As you may know, in our conventional medical system, it is often our approach to deal with these symptoms, in a sense trying to shut the body up, masking the calls for help and giving relief of the symptoms.

The cause of the distress, however, may not be addressed and therefore continues the harm. What we see, clinically, is that this can progress to more serious conditions.

I've had the good fortune to have several transatlantic conversations with Dr. Hans Scheiner, and the more I hear from him of how EMR affects the cells and stresses the body, the more it reminds me of much of the work I had been doing in emotional and physical stress.

In my previous book, *The Healthy Type A*, I developed an Early Stress Warning cue to help people tune into this stress reaction sooner, giving themselves greater success at reducing this potentially harmful response.

With Dr. Scheiner's inspiration, I began to develop an exercise to recognize the early symptoms of EMR exposure.

Radiation Rescue Early Warning Cue

This is an exercise where you take note of the first signs – the slight tingling, the slight light-headedness, etc. It is much easier to deal with these symptoms when they are just beginning, as you would assume.

This early indicator can serve as a cue, an alert to wake you up that you need to do something about where you are right now.

When you recognize this cue, the first obvious step is to reduce the exposure. For example, are you in a wireless hot spot? Has your neighbor installed a new router; have you up-graded the cordless phone model?

You might go back to the questionnaire in Step 2 and look at the symptoms you checked off. Then note the ones that seem to occur first. These can become your early warning cues.

When you recognize this cue, you know it is time to get away from what you are doing, and where you are, if you can. This is where an EMR detector is so helpful. It can alert you to high EMR areas, and also to lower ones you can seek when you recognize your cue.

Getting outside, away from electronics, is often good, unless you are close to power lines, masts or cell tower antennas.

Having a cool, or warm, shower can be helpful. And being beside a waterfall, fountain, stream, river, or ocean can be restorative.

Grounding Exercise

Next, you might consider the recommendations of some of our clinicians. One physician who specializes in energy medicine told me that he advises his electro-sensitive patients to remember the healing, restorative power of nature. This is not as far-out as you might imagine. Remember what Dr. Becker told us about the natural electro-magnetic fields of the earth.

This doctor recommends going outside and standing barefoot on the actual earth, if at all possible (sand, dirt, grass are best, but if pavement is all you've got, it'll do). Lean or crouch over so that all your fingers touch the ground. If you're feeling light-headed or dizzy, you might need support.

He suggests doing some relaxed deep-breathing, and feel as much as you can that you are reconnecting with the earth and the natural energies around you. There are probably scientific, mechanistic explanations for this grounding – might it be that with hands and feet both on the ground you're completing a circuit and allowing the buildup of artificial electro-magnetic energy in your body to be discharged into the earth?

With Dr. Gaétan Chevalier, Dr. Sinatra has written a book, *Earthing* (March 2010), on the physiological benefits of grounding. As you may know, what he calls 'good vibes' heal and balance us when we connect with the harmonious forces of the natural world – walking barefoot on the earth, for example. Yes, there is good science to back up these 'feel good' claims, as Dr. Sinatra explains:

> Our research has shown that these grounding exercises can calm and balance our systems and produce a physiological shift from the sympathetic nervous system (stress mode) to the parasympathetic (relaxation mode). The results of this shift have a positive effect on many cardiovascular risk factors: thickened blood was normalized, heart function was balanced, there was a reduction in oxidative stress from free radicals easing the disruption of blood vessels, inflammation was reduced, and there was an easing of anxiety. These are significant health benefits.

The surgeon Dr. Rea also recommends this grounding practice. I hope they inspire you to unplug more often and take better care of your body's electrical system – your heart and nervous system. This is even more critical in this wireless age. More details ahead.

As with any situation where you're feeling symptoms, it always seems beneficial to pause from what you are doing, and take a few deep breaths, letting go as you exhale. I don't mean the kind of heavy breathing that makes you hyperventilate; it's more changing the way you breathe – lower and deeper with the abdomen moving. With each inhalation, you focus on the idea that you are breathing in energizing oxygen, and with each exhalation, expelling stale air, and maybe some of your current hassles and worries along with it.

It strikes me that the ancient spiritual and health belief systems such as Taoism completely understood these principles – hence the development of acupuncture and other forms of energy medicine.

Acupuncture

Traditional Chinese Medicine (TCM) has its roots in Taoism. Its practitioners have known for at least five thousand years that the life energy of humans and other living creatures is connected with, and influenced by, the energies in our environment, whether these are natural or artificial.

TCM practitioners see the human body in terms of a system of pathways, or meridians, through which our life force energy (called qi, or chi) travels. Our health depends on the balanced and unhindered flow of this energy.

Dr. Hong Zhen Zhu, author of *Building a Jade Screen: Better Health with Chinese Medicine* (Penguin Canada 2001), has treated a number of patients with EMR-related symptoms. Although EMR is a relatively new kind of assault on our bodies' natural electromagnetic energies, Dr. Zhu applies classical TCM theory to what he's observed in his patients' symptoms.

Dr. Zhu explains:

> EMR disrupts the normal direction of energy flow along the meridians, affecting many body systems and functions. It also disturbs mental functioning, mood and spirit. Acupuncture is helpful in treating the brain fog, insomnia, and other EMR-related conditions such as Chronic Fatigue Syndrome and sensitivity to noise and light.

Dr. Zhu also recommends such mind-body balancing practices as tai chi, qi gong and meditation.

Dr. Carrie Hyman, LAc, OMD, a licensed acupuncturist and Doctor of Chinese Medicine, recommends specific herbs and increased antioxidants:

> A TCM practitioner can recommend Chinese Tonic herbs. These are a unique class of herbs that are preventative and restorative. The numerous benefits include adaptogenic properties. Many of these longevity enhancing "superior herbs", have been valued by the Chinese for thousands of years. Modern science reveals that they are high in polyphenols (green tea), flavonoids (berries...anti-inflammatory), polysaccharides and antioxidants. Moreover, they support the body's ability to manufacture endogenous antioxidants like SOD and glutathione peroxidase. This is important as EMFs increase free-radical – oxidative stress, thereby increasing our need for antioxidants.

NeuroModulation Technique, NMT

This is another treatment that works on the principles of energy medicine. NeuroModulation Technique (NMT) was developed by a chiropractor/complementary health practitioner, Dr. Leslie S. Feinberg, DC, in 2002. His website www.nmt.md describes this approach:

> The informational basis of illness and disease may be the most common, yet the most overlooked, roadblock to healing. NMT is best described as 'informational medicine,' because it works to identify and correct the informational source of illness – the confusion that can interrupt our innate healing mechanisms.
>
> Many people believe that illness results from some external affliction that viciously attacks the mind-body. The NMT position is that illness is the inevitable result of informational confusion and faults in the systems responsible for regulating body functions, making them unable to produce the balanced internal-body state that wellness requires.
>
> A good example of this is food allergy, where the immune system is so confused that the body damages itself by its misguided reaction to a food that it should see as beneficial nutrition.
>
> Seen in this light, illness or "dis-ease" in the body happens by default when the mind-body is not sufficiently aware of its internal conditions and the requirements for healing.

The mind-body is a self-correcting system that always seeks to find and maintain the balanced internal state in which optimal health and vitality thrive.

I asked Dr. Feinberg how this system might benefit people with EMR exposure symptoms and conditions:

> With thermal exposure there is obvious harm – heating of the tissue. As you have told me, most of the concern in the scientific community is with non-thermal levels from cordless and cell phones, Wi-Fi etc., where the radiation is not strong enough to heat tissue, but is causing harm nonetheless.
>
> We could use the analogy of allergy and toxicity. Mercury is highly toxic – harmful to everyone. This could be compared to thermal EMR exposure.
>
> An allergy, however, happens when the immune system is provoked to react in an 'inappropriate' way, causing reaction symptoms – inflammation etc.
>
> This could be compared to what happens on the cellular level with non-thermal EMR exposure. Perhaps it is not so much that this level of radiation is harmful to the body, as it is confusing.

Confusing? I remembered what Dr. Carlo said about the mechanism of harm being the cell membrane going into 'lockdown' mode as it does not recognize these new artificial frequencies, and that it was this reaction (misinterpretation) that affected the normal functioning of the cell.

I found this conversation with Dr. Feinberg illuminating and encouraging. You can imagine what it has been like for me to be immersed in this issue for the past three years, reading more and more research that presents increasingly troubling results.

I asked Dr. Feinberg how this 'confusion' might explain electro-sensitive people:

> It is the same with allergies. Some people are very reactive and have symptoms. With electro-sensitive people it might be tingling of the skin, headaches, difficulty sleeping etc.

Remember what Dr. Scheiner told us about people becoming "allergic" to electricity?

But what about something like leakage of the blood-brain barrier? I asked Dr. Feinberg. Although this specific area has not been the main focus of Dr. Feinberg's work, and he was not familiar with Dr. Salford's research, he replied that this could be explained by inflammation – one of the body's primary reactions.

I wanted to three-way Professor Salford into the conversation right away, and to bring in all our other experts. Hopefully, one day we can get these researchers and clinicians together to provide a more complete picture of how this all works, and how best to diagnose and treat EMR exposure. Not only for the millions of electro-sensitive people, but for all those with high EMR exposure who have not yet developed symptoms.

NMT & Autism

Dr. Feinberg told me that NMT was used recently in a research study with children with autism, led by Dr. Robert Weiner, a

Clinical Psychologist in Dallas, Texas. The trial was stopped after 18 children completed the study.

The results were so positive that Dr. Weiner wanted to report the findings as soon as possible rather than waiting another 6-12 months to collect more data before he announced the findings and published the study.

Dr. Weiner designed the study to determine the effectiveness of NMT in improving the functioning of children with autism. The study ran for 12 sessions over 6 weeks. Statistically significant improvement was noted by the end of 6 sessions, with continued improvement occurring through the end of the study.

Parents whose children were in the study were amazed at the changes they saw in their children during the course of the study. Each session averaged 45 minutes, so the study demonstrated that NMT was able to produce profound changes in these children in a total of only 9 hours of NMT treatment.

This is a small study, with encouraging results. We need to help these families get their children back so they can all thrive.

I wanted to call Dr. Klinghardt right away to tell him this news. Then Dr. Feinberg told me that Dr. Klinghardt knows his work, so is probably already aware of this study and its promising results.

I was "connecting the dots" during this conversation and saw a comparison, not only with allergies, but with the "miscommunication" of the fight or flight response – the source of our emotional stress. This is when the body gets all geared up – muscles tighten, blood pressure goes up, etc. – with our basic instincts to fight that tiger, when the only threat is slow traffic.

Another Key Factor: Acid-Alkaline – Your pH Balance

There is a Swiss naturopathic doctor who specializes in this area. Dr. Christopher Vasey ND, in his book, *The Acid-Alkaline Diet*, explains why this balance is so important to overall health, and why most of us are overly acidic. Not good news: most of the foods in our Western diet are highly acidic – meats, starchy foods (white-floured pasta), sugars (most fruits, all cakes, cookies, sodas etc.) alcohol and, of course, vinegar. Dr. Vasey reports that fatigue is a primary symptom of being overly acidic.

Other symptoms include: irritability, depression, headaches, inflamed gums, skin inflammation, red blotches, itching, heartburn and other stomach and digestive problems, excessive urination, fungal diseases (yeasts thrive in the acidic environment), joint and muscle pain and our old friend insomnia.

This list reminds me of the Chronic Fatigue and the EMR checklists. It may seem puzzling; however, everything is interconnected in our body. And when the natural balance of this vital pH is thrown off, it affects all the fluids and systems in the body, producing many symptoms.

Briefly: Dr. Vasey describes effective ways to test urine and saliva – good indicators – and how to bring the pH back into healthy balance. All green and colored vegetables – without any sweet or sour sauces – are the primary alkalizing foods and should be included in every main meal, he recommends.

This doesn't mean we give up acidic foods like meat, sugar and many grains; it does mean that we balance with alkalizing foods: vegetables, green drinks, soft cheeses, almonds and avocados.

Keeping Hydrated

This is vital. Virtually every complementary medicine practitioner highly recommends that we drink lots of pure water.

Sorry, this does not include coffee, tea, juice, soda or alcohol. This means good, filtered H_2O.

Every cell and system in the body must have sufficient toxin-free water to maintain healthy functions.

And this is crucial for any kind of detoxification. It is also helpful in terms of our EMR exposure, as much of our cells' composition, and our cell membranes, are lipids and liquid and need to stay permeable, as Dr. Teitelbaum explained.

Here's a good tip: drink your large glass of water, with alkalizing greens, between meals. There are also mineral drops and powders that you can add to alkalize your system.

It is best to hydrate and alkalize *between* meals, not with them, as you don't want to dilute your digestive enzymes. It is also good *not* to drink ice water with meals, as this hampers digestion.

Techno Speed, Stress and Everyday Frenzy

Are you discovering there are many facets to this discussion, not just health concerns? All these electro-seductions can grab our attention by the throat and not let go. I read that when we are staring at a computer screen, we blink a fraction of the times we should. Ever felt dizzy at the keyboard when you realize you have been so ramrod focused that you forgot to breathe?

We may tell ourselves, "take breaks, breathe, stretch", yet the hours go by and we're hunched up, tightened up and bleary-eyed, still staring at the screen. It's as though we've lost touch with the fact that we have a body, until we notice the aching low back, sore neck and shoulders and wrist pain.

Some people even think computering can be good down time.

The hyper-speed that all this technology affords us is a two-edged convenience. People used to have to wait for you to return messages; now, they demand instant access.

And there used to be quiet moments in our day – waiting for the kids after school, waiting for a bus, or just hanging out. Don't see much of that these days.

Now, we never have to be alone. There never has to be a gap. We can fill every bit of space with texting, if we can't be talking or shopping online. This can lead to a very speedy mind.

Our dazzling display of technology feeds right into, we could almost say enables, unhealthy stress.

Remember Toad of Toad Hall, in *The Wind In The Willows?* How he would have loved instant messaging and video gaming. Zooming from one call to the next can feel so exhilarating. We hardly need to take a breath in between.

All this speed can make us feel so alive; it is, however, hard on the system. When the body responds with a surge of adrenaline, heart rate and rhythm are disrupted, blood pressure goes up, the blood thickens – not a good combination – muscles tighten, the normal functioning of immune and gastro-intestinal systems is suppressed, and more.

Foolishly, the body thinks these responses are going to save us from that saber-toothed tiger, when in fact, these 'adaptations' work against us. (Reminds me of the cell membrane's misunderstanding of non-thermal EMR exposure.)

This misplaced stress response disturbs our sleep, makes us tense, ages us more rapidly and can lead to a wide range of symptoms and conditions. You know this feeling – you're restless, edgy and impatient. It's not only the jet-fuel java accelerating your heart, it is also the stress response.

Our high-tech toys feed our Type A need for speed. I know this from my research, and from my own personal experience: high-energy go-getters can be a tad impatient. Knowing about high-tech health risks, I realize that this personality style plays a role here.

If we are going to reduce our EMR exposure, unhook ourselves from our tech obsession, it seems we need to cut this cycle of urgency, speed and stress. A challenge: do you know anyone who is not stressed these days?

Even vacations don't seem to provide much of a gap. Here's a good space-filler I saw recently: a luxury resort is now offering your own wireless television set, complete with wireless headsets, poolside. (No, I don't remember the name of it.)

We can quite literally, go into withdrawal when someone, some thing attempts to take away our 'drug'. Try taking your child off computer games 'cold turkey', or forgoing texting, emailing, YouTubing, Facebooking and web-surfing for a day. We may need to come off slowly.

Whatever you can do to reduce the adrenaline will help. In fact, cutting back on caffeine – gradually, to avoid a withdrawal headache – is a very helpful way to slow you down and will make it easier to cut back on your tech time, assuming this is what you want to do.

Before you got caught up in this swirl, do you remember what more slow-paced pastimes you enjoyed?

Reading? Walking in the park? Sitting on a log at the beach? Without headphones.

Maybe visit a friend. Yes, actually, go and *see* them. And sit and chat. Reconnect.

When we allow our mind to slow down, our body will follow.

There are some ways that we can enjoy these together. Meditation, for example.

Just plain sitting.

Being with your breath.

Being with your mind. Being where you are. Unplugged.

An outrageous thought, perhaps?

To sit quietly. To tune in.

To just take some down time to be with ourselves, to be with this moment, somehow goes against the grain in our modern world.

Calming The Scattered Mind

Interestingly, in Japan, one of the most frenetic cultures on this planet, many people still seek out the quiet contemplative environment of Zen meditation. Nothing religious, just taking time to slow down, and tune in.

I have never been to Japan but have studied meditation, flower arranging and Tea Ceremony for many years – quite an accomplishment for this go-getter.

When I had a clinical practice and taught people progressive relaxation, I used to combine this with simple breath meditation.

This was before we were so crazed and dazed by technology, so it is even more useful now to pull ourselves away from the screen, power off, and sit quietly for a few moments.

Dr. Jon Kabat-Zinn is the author of *Wherever You Go, There You Are: Mindfulness Meditation In Everyday Life* and founder of the Stress Reduction Clinic at the University of Massachusetts Medical Centre.

Dr. Kabat-Zinn describes what it is like to have a meditation practice:

> ... by making some time for this each day doesn't mean that you won't be able to think any more, or that you can't run around, or get things done.
>
> It means that you are more likely to know what you are doing because you have stopped for awhile and watched, listened, understood.

You can try this mindfulness meditation practice in a quiet place, at any time of day.

The Posture: Sit comfortably on a straight-backed chair or on a cushion on the floor. Let your shoulders relax and let your arms hang straight down by your sides. Rest your hands, palms down, comfortably on your legs.

The Eyes: In this style of meditation you sit with your eyes open – helpful in keeping from dozing off. Your gaze is soft and relaxed; you look out and lower your gaze down slightly.

The Breath: try to refrain from any fancy breathing. You are not trying to bliss out, or get caught up in any other entertainment. Just sit quietly and breathe normally. Once you settle a bit your breathing will become more natural. Just breathe in through your nose, and out normally through your nose and mouth. Let your jaw relax.

The Attention: The challenge here is that the more you need to do this, the more difficult it may be. Some of us haven't sat quietly with ourselves for a long time.

It's hard to grasp but the action here is to just be. Just let your attention rest on your breathing.

YOU ARE NOT TRYING TO STOP YOUR THOUGHTS!

This only creates more. Let the thoughts – and all those colorful emotions – come and go. When you wake up and realize that you have been carried away by them; come back. Come back to the gentle movement of the breath.

This helps to train, focus and calm the mind.

The Benefit: What has this got to do with our Radiation Rescue? When we are no longer uneasy in the space of things as they are; no longer edgy when alone, or frantic when nothing is 'happening', we find ourselves spending more time away from the computer and our other diversions.

And every moment we are unplugged reduces our EMR exposure. When we get outside, just being in nature, our energy soars.

Yoga, Tai Chi and other Body/Mind Practices

The benefits with these practices are numerous, in this context they help us reconnect with ourselves, ease the stress in our bodies, clear the clutter from our minds and release the stale breath from our lungs, as well as focus the mind. I'm sure you can find a good class at your local recreation center, if you want to try out these body/mind exercises.

These flowing activities, particularly when done out in the fresh air, and when we breathe deeply, helps to recharge our life force energy and helps us to reconnect with the natural energy fields of the earth.

Walk, Stretch, Swim, Enjoy

Remember the River of Health, and how it can be flowing or stagnant?

An easy way to diagnose your own inner river is: if you feel tired and low energy, the flow is sluggish, toxins are building up, and you're lacking oxygen.

One of the best ways to recharge our EMR-overloaded systems is to get this river moving again; our blood moving, lymph flowing and chi circulating.

Stretching at the keyboard is good for crunched muscles, but it's even better to get away from it.

You may have built up resistance; exercise being a way to lose weight. You may have come to resent feeling pressured into being fit.

But let's look at being more active from a different perspective, one that is pivotal to staying healthy amidst environmental pollutants.

You don't have to walk, cycle or run for miles; just get a stagnant system moving. Flowing. The vitality of our lymph system is a key element in dealing with almost all of the symptoms and conditions we've been discussing.

If this clearing house is not functioning well there is a build-up of toxins, the immune system becomes overwhelmed and we are more likely to be sensitive to chemical and electro-pollution, have allergies and food sensitivities, among other problems.

Being more physically active helps the lymphatic fluid clear toxins and this improves so many conditions.

A spry 90 year-old woman at our rec center coached me on her swim routine: warm up your muscles in the hot tub, take a brief cool shower, then swim. She does ten slow and steady laps, while I splash around in the aqua-fit class. After you get out of the pool, she suggests, have a steam to help the body flush out any toxins from the pool water. Some gentle stretching and re-hydrating with lots of pure water completes her routine. Try it.

When you recognize the Early Warning Cue which tells you that your stress, or EMR exposure, levels are rising, get up and get outside. Lie on the grass. Look off into the distance.

Watch birds in the sky.

This is good for eyes that spend long hours staring at various electronic screens. Swing your arms. Breathe deeply. Move.

This, and staying hydrated, are the best ways to keep the river flowing. Swim. Play tag with your children at the park. Put on your era's favorite music and dance just like you used to do. That'll make them smile.

I encourage you to add to your action plan twenty minutes daily of being outside in the fresh air. Feel too tired? All the more reason to get moving. You may find it more effective, and more fun, than the usual ways we try to deal with fatigue – the sugary carb, or the caffeine kick. (Isn't it annoying that many of the things that taste good aren't good for us?)

The Communication Challenge

Recognize this eye roll? Persuading supposedly rational adults to adopt a lower-tech lifestyle is one thing. How do you get your teens to understand that their beloved cell phone, so central to their social life, is as much of a safety issue as sex, drug use, wearing a helmet, driving safely and so on?

I spoke with Dr. Allison Rees, a psychologist and parenting expert, who agrees that dealing with this hyper-tech generation can be a real challenge:

> Teens need and want to make their own decisions, just at a time when they aren't good at doing that! So we attempt to educate, state our concerns and negotiate, as best we can.

> I suggest not indulging children with the technology, and not seeing these devices as a privilege. I'm concerned with the sense of entitlement kids have around getting this 'stuff'.

Dr. Rees offers some ideas for communicating more effectively with your children, which may be helpful whether you're talking about technology, or any other issues.

Here are some of her key points:

- Say what you want to say; don't just react to the child's statements or behavior;

- Stay on one topic, be brief and to the point, and don't bring up issues from the past;

- Do respect the teen's need for independence and request a positive change;

- Don't lecture or try to control by making demands, commands, threats, or guilt-inducing statements, or fear-inducing warnings.

Dr. Rees adds:

> People, including children, tend to resist being pushed, coerced, or otherwise forced into changing their behavior.
>
> When a parent forces a child to change his or her behavior, the child reacts like someone whose head is held under water by a bigger, more powerful person. He/she does anything possible to resist.

If the parent succeeds, the child feels resentful and wants revenge.

It is therefore important for us as parents to make requests rather than give orders. We should recognize that it's up to our children to control and change their behavior, rather than trying to impose such changes on them.

Only in situations where there is actual physical danger should we try to control our children.

I would suggest that in the case of radiation exposure, as with other health and safety issues such as alcohol consumption and driving, you engage your teens in a problem-solving discussion around solutions that will keep them safer while still respecting their need for autonomy and some privacy.

For instance, can they agree to abide by some of the safer use guidelines for cell phones – limiting use, keeping them powered off and away from their heads at night, and using the speakerphone or wired headset?

As with so many issues that come up as our children grow into young adults, we have to be clear about our reasons and give them the information they need. As Dr. Rees reminds us, lecturing breeds resistance.

Other Technology Issues:

A 2009 Good Morning America segment interviewed five top texting teens – one boy sent up to 35,000 messages a month. I was amazed that his parents did not seem horrified with this saying, "Well, it doesn't seem to be affecting his school work".

Let's consider the radiation exposure. However, there are concerns with teens and technology that go beyond physical health effects and distractions at school. For instance, vicious attacks of cyber-bullying, where a young teen is targeted with abusive emails or instant messages.

Or the obsession with texting which has developed a sideline where young people send each other nude photos of themselves by cell phone. You may want to check in with your children and teens about their communications. Check out the bill to see how much the phone is being used.

Internet Safety

If our children were confronted on the street, or in a video arcade, by the stranger they may be freely opening up to online, they might be horrified. And, perhaps, in danger. The health hazards of wireless radiation are not our only concern with all those hours spent online. No, it's not all homework. Ever notice that they suddenly switch screens when you walk in?

Last night we heard an Internet safety expert talk at our children's school – an eye opener for the parents and students.

I asked Jesse Miller, founder and lead at Miller Consulting Services, for his insights:

> I am concerned about users who participate without regard for privacy and the personal repercussions of having an open profile. Everything you post: messages, pictures, videos etc. become the property of that website. And social networking websites are actively providing your personal information to other companies, with your approval based on agreement of the prescribed terms of use.
>
> Not knowing, and/or not caring, about any consequences, our youth see Facebook, for example, as a place where they can be who they are, communicate with friends openly, without adult judgement and make certain to stay in the loop about personal and world events.
>
> The consequences, however, to this behaviour when un-monitored, or addressed by parents, can be detrimental to their endeavours in the future. Most

young people, and adults, do not know that permanent traces are left behind. It is common for potential employers, school and university admissions to use Facebook, and other sites, in their screening process. The young people, and adults, I talk with assure me that their profile is closed and secure. This is not always the case and I show them how easy it is to access and retain their personal information, pictures etc.

Less common, but potentially serious and tragic, police investigations track home invasions, abductions, sexual assaults and homicides that have resulted from seemingly innocent online encounters. I encourage parents to explain to their children that their passwords should be available (maybe in a sealed envelope) as access to these accounts will be one of the first steps in a police investigation.

I advise parents to discuss online activities openly with their children and realize that what youth communicate in this way is no better and no worse than what previous generations did when their parents weren't looking, or listening.

The difference is that today this personal information – including pictures, schools, home addresses, when you are home alone, away on holiday, and telephone numbers – is digitized through social networking multimedia and in turn being brought into public view, to anyone seeking it.

The Internet has many benefits, as we know. We do, however, need to know how to use this resource more safely, and securely. (More details at www.howsafe.ca)

The Low-Tech Family Challenge

Roberto Salti and colleagues at the University of Florence in Italy found that when 74 children, aged 6 to 12, were deprived of their TVs and computers for a week, and other sources of artificial light were reduced in their homes, there was a beneficial outcome: the production of melatonin increased by an average of 30 per cent. The increases were highest in the youngest children.

And as you may know, being outside in the sunshine triggers the internal production of Vitamin D, which is also essential to our health.

Ever wondered what might happen if your family were to unplug from the virtual world and reconnect? This is for families wired on wireless who want to reconnect with each other, and the natural world around us.

If you've had children in primary schools, you might have participated in "turn off the TV week", when families are encouraged to spend time reading, playing games, listening to music, going outside – any activity that doesn't require a television or computer monitor.

There was a TV challenge some time ago for families to go without for a month. I remember one young teen complaining it "was like a death in the family". Interestingly, there were similar stages – irritation, boredom, anger and eventually, if they could hang in there, some times of once-lost fun and jokes.

Some parents reported that it was really refreshing not to have to drag the kids away from the TV show or the "next level" in the video game when it's time for dinner.

I'd like to propose a similar experiment: a trial run at a lower-tech family life. Remember the joy of reading, books, the printed page?

Don't panic – you don't have to toss out all things electronic. I'm hoping this process will give you the time and mental space to see how much of the technology is really necessary in helping you function and stay connected, and how much is just a habit and mental clutter.

The first step might be to take another look at the survey and get a sense of just how much technology is central to your family's activities. In the evening is everyone in their own space texting, emailing, watching TV, playing video games? Isn't everyone, you might ask.

There are some families, I must admit ours is one, who enjoy some "no-screen" evenings listening to music and reading, or playing cards, board games and sharing lots of laughs.

Last Spring Break we went to a nearby family resort that offers lots of outdoor activities. We were surprised that the nature trails were empty, while the parking lot was full. We had a wonderful time exploring the woodsy wonders below the forest's canopy, digging in the inter-tidal zone at the water's edge and sitting on logs watching birds.

We have given our children second-hand digital cameras; this is a great way for them to develop an appreciation of nature, and have fun taking pictures and sharing them with their friends.

Reconnecting with each other doesn't have to be rare family holiday time, it could be part of your weekly life.

One Family's Low Tech Challenge:

Here's a follow-up with our 'library mum' and her journey with this book:

> Since reading the last parts of your book:
>
> 1. Thought about nature deprivation. We don't live near any woods but I unplugged the kids and went to the park the other day. Played and had a wonderful time.

2. Today, took the kids to the ocean...bypassed children's playground and went straight to tide pools and rocky shore. Had a great time with them exploring barnacles, shells and watching the tide...thought about nature deprivation and how wonderful it felt to be with them in this unscripted environment.

3. Thinking about how to redesign life in ways that aren't just about restricting or taking away toys/conveniences, but finding creative ways to do things that contribute to overall wellbeing...this is in response to my son who is so concerned that our discussions will lead to the restriction of things that he loves (and it probably will, but he may end up happier as we spend more time outdoors in nature together!)

Library mum signing off for today on her way to buy a battery-operated alarm clock.

It could make a powerful difference if more women were as quick to re-think their families' balance between nature and technology, as our library mum and her neighbour have.

There was a recent article from BBC news in the UK, reporting on a surging number of "geek mums".

It seems many women are using their emails, cell phones, texting, Twittering, blogging, to counter the loneliness and social isolation of being at home with children.

Consider the fact that more than 60% of electronic devices are evidently purchased by women; you see the potential impact mothers, grandmothers, sisters, aunts can have on this issue, if they choose.

It's quite amazing that a small shift in the family focus can bring great benefits. We haven't made any major changes – just less time on the computer – all of us that is, as I was spending a lot of time there myself writing this book. We are making an effort to get outside more, even in less than brilliant weather.

Is Technology Endangering Our Children's Connection With Nature?

Have you heard that the latest version of the Oxford Junior Dictionary (Oxford University Press) has deleted nature words like: beaver, dandelion, heron, magpie, otter, acorn, clover, ivy, sycamore, willow and blackberry? In their place, they have substituted more popular terms, like the electronic "Blackberry", "blog", "MP3 player", and "broadband".

Canadian wildlife artist and conservationist Robert Bateman, whose "Get to Know Program" has been inspiring children to go outdoors and "get to know" their wildlife neighbors for more than a decade said in an interview with The Canadian Press:

If you can't name things, how can you love them? And if you don't love them, then you're not going to care a hoot about protecting them or voting for issues that would protect them.

If you are feeling inspired about changing direction with the focus of your family so they will have a greater appreciation of nature, and unplug now and then, I encourage you to approach this with tact and skill.

As you know, it is not easy to guide your children in the direction you want them to grow. And it can be irritating, even heartbreaking, that they seem to insist on moving in the opposite direction – you want them to be playing outside in the fresh air, and they prefer being inside, glued to video games in darkened rooms.

Richard Louv, author of *Last Child In The Wood*s, has a great quote in his book from a child who declared, "Of course, I prefer to play inside. That's where all the electrical outlets are."

Richard shares his insights on saving our children from what he calls nature-deficit disorder:

> People around my age enjoyed a kind of free, natural play that seems, in the era of kid pagers, instant messaging, and Nintendo, like a quaint artifact.

Within the space of a few decades, the way children understand and experience nature has changed radically. Today, kids are aware of the global threats to the environment – but their physical contact, their intimacy with nature, is fading.

Yet, at the very moment that the bond is breaking between the young and the natural world, a growing body of research links our mental, physical and spiritual health directly to our association with nature – in positive ways... The health of our earth is at stake.

Richard Louv goes on to inspire us about the possibility of a "new frontier – a better way to live with nature". His work has evolved into the much-heralded Children & Nature Network (www.childrenandnature.org) and the Leave No Child Inside program.

It's tempting to keep including quotes here from his wonderfully thought-provoking *Last Child in the Woods*, Algonquin Books, 2008. This wake-up call offers so much great information that it is pivotal to our Low Tech Family Challenge.

Here's a story of a neighbor, an overwhelmed dad – who is inspired to make some changes in his family's life.

When He Was A Boy

This dad became so electro-sensitive that he sought not only to regain his own health, but to prevent such a situation with his young family:

> When I was younger we were outside most of the time, except for those long hours in school. My friends and I used to make up games where we would roar around and using our imaginations in all sorts of scenarios.
>
> We discovered how things worked by putting things together; well, mostly taking things apart in all sorts of places – in the garage, the back yard and down by the river.
>
> Some days we would just sit there and watch insects. But mostly we were climbing trees, challenging ourselves. I remember working with the fast flowing river discovering how to have fun without drowning.
>
> I didn't know it at the time, but we were developing social skills – working with each other, getting along and sorting out our differences – who was going to lead the next game.

We used to make up a lot of games – let's play ... in order to keep the game going you had to get along. We began to see other people's skills – Joey was very good at seeing things intuitively, and Billy could fix things in his dad's tool shed – we relied on him for many engineering projects.

We developed trust in each other; we were in real situations where you had to count on people.

We were looking at birds squawking at each other, all the life in the wetlands, watching the fish and their cycles. You learn a lot about life cycles watching salmon – see them spawning, and watching them die.

Without making a big deal of it, we were connecting with the impermanence of life, the changing of the seasons, and connection with how things are.

I remember bright summery days outside when my mother would be picking dandelions to make wine and we all picked blackberries – and ate a few – for her absolutely wonderful pie.

My wife and I both work hard. We put in long hours – virtually all inside – and it is a struggle just to keep the kids fed, dressed and schooled, so we're not baking any pies.

But I see that there are so many times in the day where we are making choices that involve sitting in front of a computer, TV or video screen.

Inspired by Radiation Rescue, we decided to make an

effort to wean ourselves off of these entertainments. It seemed daunting at first. The kids rebelled.

Then my wife and I sat down and had a family discussion about how all these technology things had been part of making me sick.

They, of course, shot back that was because I was 'old' and they were just kids so it was okay for them.

I didn't subject them to all the evidence in the book, but my wife and I did our best to explain why spending too much time around these things was also not a good idea for kids.

We found that rather than trying to tackle this head-on, that we would plan some outings.

As I unplugged more myself, I got better – I was sleeping much better and had more energy.

So instead of collapsing in front of the TV when I got home, I'd say 'let's go outside and kick a ball around'.

Once we were outside more, I noticed the kids were really into it and even took themselves outside now and then. Our somewhat pasty-looking kids began to look healthier.

Sometimes we just go outside and sit on the ground and chat. And listen to each other.

What Was It Like When You Were Young?

What kind of stories are the children you know going to tell? What stories, and values, are they going to pass along to their children, and grandchildren?

If the rate of technology continues at the current supersonic speed what kind of lives are all of us going to have in ten years? In twenty?

It's not too late. We can take better care of the health of our families, and our home – this blue planet is our only home.

From Good Intentions To Effective Changes

This is a lot of information; important things for us to consider to safeguard our families. The challenge: not to get overwhelmed and give up into hopelessness.

After you waded through Step 3, I included an intention plan to help focus on what you wanted to deal with first – your priorities.

Here is that list again. Just skip over it, if you want.

Reviewing Our Radiation Rescue Intentions

Top Priorities – what I want to tackle first:

_____

_____

_____

Secondary priorities – what I want to deal with after that:

_____

_____

_____

I should consult a professional to assess the following:

_____

_____

Other environmental exposures I want to address:

_____

Now that you have been through Step 4, which has enough aspirations to last a lifetime, let's look again at winnowing this down into bite size chunks.

Our Radiation Rescue Action Plan

Area	Action intended	Action completed
Front Porch		
Master Bedroom		
Children's Rooms		
Bathroom		
Kitchen		
Laundry Room		
Vacuum Cleaner		
Office		

Area	Action intended	Action completed
Cordless Phones		
Internet Access		
Other Environmental Exposures		
Your Physical Health		
Nutritional Supplementation		
Dental Health		
Other Well Being/Lifestyle Factors		
RR Family Communication Challenge		
RR Low-tech Family Challenge		
Other Personal Goals		

Checking in

Please don't imagine that you are going to turn all of this around at once. We have been developing a high-electronic environment for a long time. Let's just take this one step at a time, and feel good about each exposure we reduce. If you have got this far in Radiation Rescue, you have already accomplished a lot.

Thanks to our international group of EMR experts, you now know more about the causes, and solutions, than many of the people who are supposed to know about them.

I hope you will continue through the last chapter to hear from some of them about their hopes for the future, while you contemplate yours.

Hope For The Future

Moving from Knowledge to Action

It may surprise you to know (or maybe by now it won't surprise you) that some American scientists have been telling their government about EMR biological effects for more than 35 years!

Robert O. Becker, MD, was one of the researchers looking into the U.S. navy's plans to install a 6,000 mile underwater messaging system to communicate with submarines. Dr. Becker said:

> As far as I know, our testimony (in 1975) was the first ever openly given by American scientists stating that electromagnetic energy had health effects in doses below those needed to heat tissue, and that power lines might therefore be hazardous to human health.
>
> We criticized the White House Office of Telecommunications Policy for failing to follow up

a tentative 1971 warning that harmful effects from electro-pollution were proven.

As of the writing of this book (2009) there are hopes that the US administration will take a courageous stand on this issue. As I mentioned early on, one of the leading lights in fighting the installation of cell towers is Eileen O'Connor, the founder of Radiation Research in the UK. Her request for input to the panel studying the possible links between radiation (nuclear and electro-magnetic) and cancer has been accepted.

Ms. O'Connor is pleased the issue is being taken seriously:

> President Obama recently said science is about ensuring that facts and evidence are never twisted or obscured by politics or ideology. It's about listening to what our scientists have to say, even when it's inconvenient and I welcome this approach.

Many of us have the "audacity of hope" that President Obama and this administration will show true leadership in this other inconvenient truth. Any steps the panel recommends, and that the President takes to protect people's health, are likely to be echoed around the world.

As you may know, most politicians take their lead from the majority of the electorate's demands; it's not the other way around. Yes, there's industry lobbying, but when enough of us unite to demand safer limits for this technology for the health of all of us, especially our children, then things will change.

What Your Town Can Do – Remember Dr. Oberfeld and Salzburg?

Have some shiny new palm trees (pine trees? church spires?) sprung up in your neighborhood? Near your child's school? You can do something about this.

There is an excellent book edited by B. Blake Levitt: *Cell Towers: Wireless Convenience? or Environmental Hazard?* This is a quote from Blake Levitt:

> "Something municipalities fail to keep in mind is the basic legal fact that it is up to the providers of a service or product to prove that their wares are safe. It is not up to us to prove that they are unsafe. The tele-communications industry has largely failed to do that. Just because they are within the FCC guidelines for RF emission, does not prove safety.

> No town today should allow itself to be intimidated by telecom service providers or adjunct industries like tower companies... Those in decision-making positions need to understand that this form of land use regulation is very different than traditional forms. This is NOT just an aesthetic issue. It is a medical one."

This form could be used at a public meeting on installations of cell tower antennas (masts). Most likely none of the officials would sign it. Their reluctance, however, would be of interest to the residents who will be exposed to the radiation.

Statement of Accountability

This is to confirm my stated position that the (wireless technology, cell tower/broadcast antenna (mast)) that has been (proposed, approved) for installation in (…) is, in my informed opinion, safe. I am confident that there are no significant health risks for any adult, child, or animals/birds/bees living within _____ *(please state your view of the proven safe distance)* of this system.

I have been informed of the research indicating significant biological harm that can result from the electro-magnetic radiation emitted from this kind of system and mobile phone technology. I have read the February 2009 statement of Dr. David Gee, one of the authors of the BioInitiative Report:

> **Mobile phone manufacturers will not be able to claim immunity if, in ten years or so, it is determined that use of mobile phones causes brain cancer or other illness. They are forewarned now, and simply saying 'but we were in compliance with ICNIRP safety limits' will not be sufficient reason for ignoring the growing evidence.**

(If you have seen a signed assurance of the health safety of this system from a medical expert, with recognized expertise in the health effects of wireless communication systems, please list his or her name and credentials:

_____

The main reason(s) I have recommended this installation be implemented is:

_____

I certify that .... (name(s) of individual, company, elected officials, government agency) do (does) not have any outstanding health and/or legal (claims, disputes) regarding this technology at this time.

Please note exceptions:

_____

This is to verify that any adverse health effects attributed to this installation will be covered by our insurance policy provided by:

_____

Your name: (*Please print*)_____

Your title and official position:

_____

Today's date: _____

Your signature: _____

Witnessed by: _____

For your information, a list of the individuals who have been sent this letter has been sent to ...... (media names).

As there is strong community interest in this issue they have been sent a copy of this letter and asked to follow up with you.

(This ia available at: www.radiationrescue.org)

Professor Johansson is widely regarded as one of the leading experts. When officials, at all levels, heed his words we will all be more hopeful for the future:

> I strongly advise all governments to take the issue of electromagnetic health hazards seriously and to take action while there is still time. Governments should act decisively to protect public health by changing the exposure standards to be biologically-based, communicating the results of the independent science on this topic and aggressively researching links with a multitude of associated medical conditions.

As you know, there are many things we aspire to in this life. Fundamentally, is there anything more truly worthwhile than safeguarding our family and the future of our children?

That's why I have spent three years researching and writing this book. For you, and for your family.

Promising Signs On The Horizon

Are you old enough to remember the movie The Graduate? A young man is given a tip for success by one of his parents' friends, "The future, Benjamin, is plastics." The advice now might be: fiber optics. Why? Several of our experts have mentioned this as the wave of the future.

Progress in Technology

Fiber Optics

Many telecom technical experts also prefer this system. One told me that, "Fiber optic lines are thin strands of optically pure glass. This is the best way to move large amounts of data – digital information – over long distances.

Security is a big factor, (as we mentioned before). Fiber optics is a secure, closed system (the shotgun approach of satellites and wireless radiation firing through space is not). And fiber optics do not have the same radiation issues."

Evidently, there are plans to bring fiber optic connections to homes and businesses, instead of wireless, for our Internet access. Let's promote this. Apparently, not everyone in the wireless industry is onside.

Breaking Out of the Cyber Cocoon

If you have read through most of this book – glanced at a few pages? – you know this message is piercing.

It cuts to the heart of our techno-addictions: have we become too attached to the convenience and to those scintillating 'pings' and downloaded rings of cyber-connections? Are we immersing ourselves in clouds of radiation in the process? Are we closeting ourselves, and our children, inside and filling space with hyper-speed entertainment, the instant gratification of texting and Twittering?

Acknowledging any addiction is a tough message to take in. And that's without all of the "too scary" stuff about cancer and brain damage.

I have tried my best, without sugar-coating anything, to focus on the ways we can turn around this potentially perilous situation, and reduce the health risks, without losing the conveniences of this never-ending stream of technology.

Now, let's rouse our energies and look on the bright side of possibilities and prevention. Working together, let's do what we can to rescue our families. I am hopeful.

As a whole generation of baby boomers age – many who are already self-proclaimed 'crackberry' addicts – there is concern about health effects after so many years of sitting in front of computers and being hooked on electronics.

It is encouraging that there are treatments evolving to deal with these effects.

Clinical Progress

There is hope on the clinical horizon. Innovative doctors are working on effective ways to heal, and prevent, symptoms and conditions related to EMR exposure. There is also hope for the millions of people already suffering from electro-sensitivity, and for those with Autism Spectrum symptoms and related neurological conditions.

Let's demand more research in this area as the conventional medical approach of relying mainly on X-ray diagnostic tests, pharmaceuticals and/or surgery does not seem the best approach when dealing with the environmental assaults which we have been discussing.

It seems we need a multi-disciplinary approach focusing on holistic, non-invasive ways to heal and prevent environmental conditions. And, as the electro-magnetic fields affect us on a frequency level, we should be looking at energy medicine for some of the clinical clues.

I am going to continue calling upon our clinical and research experts, and others, to give us the information and tools we need to safeguard our health.

Parents and Grandparents Demanding Safer Products

I am confident that most people will follow the lead of our library mum and make an effort to cut back on cell phone use and get our families unplugged and outside whenever we can, and that informed parents, grandparents and educators will limit our children's exposure in the ways we have discussed.

As we have seen in Europe, legislators are listening to the independent scientists and inching forward to bring regulations within recommended safe limits. All are encouraging steps. Will all of this be in time to prevent the global EMR-related pandemic that some fear? Can we rescue this hyper-tech generation? Yes, if we act now.

Progress With Our Families' Low-Tech Challenge

Library Mum

> Basically, *Radiation Rescue* has made me more aware of the addiction that my family and I have nurtured around technology. I will be increasingly more wary of gadgets which promise to increase my well being by providing me with efficiency tools.

In his video series, John Bradshaw uses the metaphor of a hanging mobile to show that when one part of the mobile moves and changes, it sends reverberations and changes throughout the whole system. I feel that this book has done a similar thing with our family. It seems that the changes that I am attempting to implement are leading to profound changes in our family interactions in general. It hasn't all been easy... at times. On one hand, these changes have been bringing tensions to the surface (for example, my son's Internet and technology savvy seem to be in direct conflict with my desire for a low radiation home and a de-emphasis on 'screen culture').

So, through these changes, we are learning how to navigate each other's needs and respect each other; whereas without this particular technology challenge, we have had very little conflict to work through (I found myself at the height of frustration yesterday when I thought that my son was reconnecting the Wi-Fi and I pulled the computer power cord out of the outlet and banned the computer from use!)

The changes that I've been making have also had an impact on my relationship with myself. The cell phone, in particular, makes me feel important and necessary; efficient, productive and connected.

This is a heady mixture in a society that places such value on these qualities. The cell phone is a transformative wand which takes me from being a 'housewife' and 'mother' (negative terms in today's society) and magically transforms me into a 'modern, West Coast, work-at-home executive.'

Is this true? Is it what I really feel and think about – or is it the effective machinations of savvy marketers?

Because every change invokes a challenge to a personal relationship, my attempts to cut down the electromagnetic radiation in our lives are not simply a matter of just making simple technology changes.

For this reason, I have experienced profound shifts. All this navigating takes personal energy.

And so, I have not been able to implement all of the book's suggestions in a straightforward, progressive way.

Instead, the process looks messy – it has turnarounds and circle-backs. This will likely continue. Yet, as I reflect on this today, I believe that this is natural.

Even blame, self-doubt, perfectionism and discouragement can be seen as part of a larger spiritual process of growth and renewal (as long as they don't lead to an abandonment of the whole project!).

We begin to exercise our right to make choices and to create an environment rather than accepting one that is pre-packaged. This, in itself, is sufficient reason to begin the journey!

Neighbor Dad

Changing gears is not as easy as it may sound. Some days, I feel we have failed hopelessly with our good intentions to live a more simple life, and to rely less on all this technology. The least bit of rain keeps the kids inside and begging for their gaming. My wife still goes right to the computer in her jammies, bleary-eyed and clutching her coffee mug, to check her email. We have made some progress since reading *Radiation Rescue,* we should feel good about that, but this is going to take some time. I'm determined, though, to ease us gently away from this fixation, and focus on the successes, and more family fun, which we are enjoying these days.

Your Family Action Plan?

As you can see, this is a process with "turnarounds and circle-backs". As these families have described, we are not only dealing with our own households, we are dealing with a culture that has evolved around instant gratification. Please take it step by step, feel good about every exposure you reduce, and don't give up.

Your Low-Tech Family Challenge?

When you are ready, see if you can get everyone on board and make this a sporting event. Try a positive approach: add more of the outdoor activities that your family enjoys, rather than ripping the gaming stations out of the kids' hands. And, please be patient, with your family, and yourself.

Each week, spend one evening away from the computers and the TV's, and get outside for a walk in your neighbourhood, or a nearby park. Try "no-screens" for a day. Or a week?

Check out garage sales for good deals on bikes, balls, badminton racquets, playing cards and board games. Ask everyone in your family, depending on his or her age, to come up with a non-electronic activity for that week. Maybe it's a rousing game of cards, or charades?

I know we are really busy, and many of our children have back-to-back lessons, sports and other after-school activities. These too can contribute to limiting EMR exposure, particularly if we watch the soccer game instead of texting on the sidelines.

I read one local hockey arena went wireless so the parents could work on their laptops during those long games and practices.

How often do you see families at a park, or on the beach, and instead of being together, instead of really being there, they are immersed in the virtual world? What does this teach our children about how to interact in the real world?

Once you go through the withdrawal and your texting, gaming, Twittering fingers stop shaking, you may find you feel greatly invigorated by low-tech activities.

This technology, and the world-wide Internet, have brought our global village many benefits. We can continue to enjoy and expand upon those. As with most powerful influences, we just need to use them more wisely.

Progress In Consumer Awareness

The message is getting out. People are waking up. Remember that great line from the film Network? "We're mad as hell and we aren't going to take it anymore."

Dr. McKinney has offered to work with me to develop awareness and education and programs to get this message out:

> As results surface from scientists, researchers and clinicians it becomes important to share this information with citizens around the globe.
>
> Designing educational programs for parents, teachers, political representatives, corporate executives and healthcare professionals is a necessary and progressive step to increase awareness. The information within such programs should center around scientific results. We must emphasize the

positive steps people can take, at all levels, to bring about the change we need. The intention is to raise awareness for informed decision making, not to invoke fear.

We will keep this updated on our website. And will continue to bring you the latest research and technical advice.

We owe a debt of gratitude to the many people who helped with this book and to the thousands more working tirelessly in this field.

Recently, I wrote to one of our contributors:

> Thanks to the dedication of pioneers like yourself, I predict we are reaching a tipping point where consumers, particularly parents and grandparents, are waking up.
>
> Soon the industry is not going to be able to hold back the tide of public demand for the truth, for truly safe products. And governments will be pressured into action for accurate safe standards.
>
> This book, hopefully, will be a contribution to this wireless wake-up call. I am deeply grateful for your work, and all the help you are giving to these readers.

I have asked some of this group of contributors for their hopes for the future.

Dr. William Rea, and several others, offered that they were optimistic. This is encouraging as Dr. Rea, for example, is treating people every day who are already suffering from EMR exposure.

Alasdair Philips of Powerwatch:

My hope for the future is that politicians and business leaders start acting with integrity regarding sustainable living and deal with problems like EMF/RF exposure now instead of ignoring them in order to make short-term financial gains.

Dr. Magda Havas:

I hope that the value we place on the health of the environment and human health for all, including the next generation, will prevail over the greed of a few.

I hope that those individuals who are in positions of authority, and this includes parents, teachers, employers, as well as government officials, will make the necessary changes required to minimize our exposure to electromagnetic pollution.

I hope that it will not take decades, as it did with tobacco, to do the right thing.

May those who are sensitive, who are the 'early warning indicators', not suffer in vain.

May their experience alert us to the problem and give us the courage to resolve it.

May we have the wisdom to live up to our name Homo sapien sapien and not have to change it to Homo greedicus ignoramus.

Dr. Sarah Starkey:

I would like to see the Salzburg limits introduced everywhere (0.06V/m for outdoor exposure and 0.02V/m for indoor exposure for radiofrequency radiation.)

My hopes for the future? That people will come to accept that creating a sustainable future for our planet involves reducing artificial electromagnetic fields.

I would like children to grow up knowing that to protect their own health, that of future generations and their environment they need to limit their use of microwave-emitting technologies.

Dr. Heather McKinney:

We didn't get here overnight, and we will not remedy this overnight. Yet we have to start somewhere and work together. Every step counts.

Making change may seem daunting, yet at what point does the precautionary principle outweigh the cost of continuing this experiment?

My hope for the future is that consumers around the world will raise their voices to encourage scientists, politicians, and corporate executives to be proactive, and look at this situation as an opportunity for research and long term restructuring.

I am optimistic that this can happen, as long as people heed this wake-up call and realize what they need to do to protect their families.

Dr. Carrie Hyman:

We need more private citizens asking questions about the safety of these products. We all need to work hand in hand with clinicians, environmentalists and politicians to address this health issue quickly and responsibly.

My hope is that more people will take political action, e.g., write to your local and national representatives demanding your government adheres to the emission standards and recommendations set forth by the Bionitiative Report. We also need more research into safer wired and wireless devices, and science-based intervention technologies to mitigate the harmful effects.

In the meantime, I hope that parents will limit their own use of cell phones, and that of their children. Now most European countries, Russia, and India recommend that children under 16 simply do not use cell phones...except for emergencies.

Dr. Louis Slesin:

My hope for the future is some breaking down of the research roadblocks. This is a challenge as there are so many vested interests on the hazards side; so much money involved.

To be honest, I am not optimistic. I am encouraged, however, on the clinical side; there's progress with biomedical applications – this may be the key.

Dr. Devra Davis:

American children and their parents are not provided the basic protections afforded their European counterparts and are asked simply to have faith that everything is all right. Faith may move mountains, but it is not the way to protect public health.

The government should work with phone companies to protect our children and to provide data to Professor Gandhi, and to other independent researchers, so that we can find the safest way to use this revolutionary technology which has also brought many benefits.

In an ideal world, the profitable companies would compete not just to promote sports (many facilities are sponsored by wireless companies) but to design the safest sleekest phones.

I share Professor Johansson's view and also hope that I am 100% wrong about the serious public health issue we are going to face, if we don't act soon.

My hope is that as more and more people know about this issue we will have the scientific experiments we need to learn what we need to do to protect ourselves, our children and our grandchildren.

In conclusion, may I offer my hope for the future: that we wake up and demand that our health and the wellbeing of our children, our ecosystem and all life forms on this planet are not endangered for corporate – and government – profits.

I have been watching with growing concern the vast four-pointed chasm, and lack of communication, between:

1. the independent scientists in the know;
2. the bureaucrats not keeping up with the science;
3. the industry whirring out wireless hazards – on earth and in the sky;
4. and us – the misinformed consumers.

Don't lose heart; this gap is narrowing.

There are indications of progress around the Western world, particularly in Canada, the USA, the UK, Germany, Austria and France.

As Dr. Davis has said, there are benefits to this technology, and it's not going away. Let's learn how to use it more safely, so we can enjoy the convenience without the risks.

I feel confident in speaking for our contributing experts when I tell you how much we hope you hear our Radiation Rescue message and do everything you can to safeguard your family.

This will bring benefits, particularly to the wellbeing of this hyper-tech generation.

May we rediscover the delights of stepping outside of our cocoons.

May we unplug and reconnect with each other and the natural world around us.

May you, and all of our children, and grandchildren, thrive.

And may we have the courage and discipline to face the inconvenient truths, for the wellbeing of our families and the earth, this blue planet – our only home.

> *You will succeed if you persevere; and you will find a joy in overcoming obstacles. I have confidence in the human spirit.*
> Helen Keller

Resources

Research, Advocacy and Electro-Sensitivity Support Groups

The USA

• www.antennafreeunion.org

The San Francisco Neighborhood Antenna-Free Union (SNAFU) is a grass-roots, city-wide coalition of individual residents and neighborhood organizations that works to prevent the placement of wireless antennas on or near residences, schools, health care centers, day care centers, senior centers, playgrounds, places of worship, and other inappropriate locations.

• www.bioinitiative.org

In 2007, an international working group of scientists, researchers and public health policy professionals (The BioInitiative Working Group) released its report on electromagnetic fields (EMF) and health. The group documented serious scientific concerns about current limits regulating how much EMF is allowable from power lines, cell phones, and many other sources of EMF exposure in daily life. The report concludes the existing standards for public safety are inadequate to protect public health. See the *BioInitiative Report: A Rationale for a Biologically-based Public Exposure Standard for Electromagnetic Fields (ELF and RF)*.

• www.electromagnetichealth.org

One of the most effective voices is that of Camilla Rees. She is the founder of ElectroMagnetic Health and co-author with Dr. Magda Havas of the book, *Public Health SOS: The Shadow Side of the Wireless Revolution* – available on the website. On the website there is also a petition urging government action.

• www.emrpolicy.org

The mission of The EMR Policy Institute is to foster a better understanding of the environmental and human biological effects from such exposures. Its goal is to work at the federal, state and international levels to foster appropriate, unbiased research and to create better cooperation between federal regulatory agencies with a stake in public health in order to mitigate unnecessary exposures that may be deemed to be hazardous.

• www.energyfields.org

The Council on Wireless Technology Impacts is concerned about the safe use of technologies which use electromagnetic radiation, especially where we live, where our children attend school, where we work and where we pray. CWTI is a federally recognized, tax deductible 501(c)(3) nonprofit charitable organization, registered as a non-profit corporation in California.

• www.ewg.org

Environmental Working Group, a Washington, DC - based nonprofit organization, is dedicated to protecting children, babies, and infants in the womb from health problems attributed to a wide array of toxic contaminants. Their other mission is the relevant federal policies.

• www.guineapigsrus.org

Joanne Mueller's pamphlet, *Are You A Guinea Pig?* tells how to sleep better and feel better by reducing EMR exposure. You can order this pamphlet from the website.

• www.microwavenews.com

Microwave News is meticulously researched and documented, making it one of the world's most authoritative sources of information on EMR/EMF health risks. It began as a newsletter in 1980 and went online in 2003. Its founder, and editor, Dr. Louis Slesin, has worked tirelessly on this issue for nearly 30 years.

• www.momsforsafewireless.org

Christine Hoch founded Moms For Safe Wireless, a national non-profit organization dedicated to enhancing the public's understanding of wireless product safety, especially cautioning parents about children's use of cell phones. Based in Virginia, the group engages in educational activities aimed at increasing awareness of radio frequency radiation, electromagnetic radiation, and the potential health implications of wireless products and technology.

CANADA

• www.hans.org

The Health Action Network is a charitable natural health resource. HANS provides information on preventive medicine and natural therapeutics through its website, a reference library, Health Action magazine, the HANS e-News, and regular events.

• www.weepinitiative.org

The Canadian Initiative to Stop Wireless, Electric, and Electromagnetic Pollution (WEEP) has a mission to inform the public about the potential environmental effects associated with various forms of electric and electromagnetic immersions.

IRELAND

• www.ideaireland.org

The Irish Doctors' Environmental Association (IDEA) is an organisation established by health professionals seeking to promote the right to health and peaceful co-existence worldwide.

UK

• www.es-uk.info

Sarah Dacre is a trustee of ES-UK. This is a registered charity in the UK advising and educating those affected and the wider population about electrosensitivity, its possible triggers and ways to lessen symptoms. Their hardcopy newsletter is sent out to ensure it reaches those unable to access information on-line.

• www.mastsanity.org

Mast Sanity is a national group of community members who campaign against the unsafe placement of mobile communications towers.

• www.mast-victims.org

This is an international community website for people suffering adverse health effects from mobile phone masts in the vicinity of their homes. The purpose of this website is to bring together people from all over the world who have become victims of insensitive mast and antenna siting.

• www.powerwatch.org.uk

Powerwatch is a non-profit independent organization that has been studying the health effects of electro-magnetic fields for the last 20 years, and provides information to help the public understand this complex issue. It also works closely with decision-makers in government and business, and with other like-minded groups, promoting policies for a safer environment

• www.radiationresearch.org

EM Radiation Research Trust researches EMR and related health effects; has a leaflet called *Precautionary Approach to Wireless Communication;* lobbies the U.K. Government and the E.U. to fund and prioritize the research effort and campaigns for law and policy change based on proven science.

• www.wifiinschools.org.uk

Wi-Fi In Schools has been set up by Dr. Sarah Starkey and a small group of other scientists concerned about the rapid spread of wireless technologies in schools. The group recommends application of the precautionary principle until better long-term health studies have been carried out.

• www.wiredchild.org

WiredChild is a charity run by a group of concerned parents, raising awareness of the potential risks to children of exposure to radiation from mobile phones and other wireless technology. Their hope is to enable other parents, schools and children achieve a better balance between safety and convenience.

EUROPE:

• www.hese-project.org (German)
• www.hese-project.org/hese-uk (English)

HESE (Human Ecological Social Economical project) is an international, interdisciplinary collaboration of scientists, institutions and informed laypeople, which brings together research and advocacy information in several languages. Many of its contributors have suffered health effects from EMR. From the English-language website, there is also a link to a new Swedish group's EMR education project, "Faraday's House". This radiation-shielding structure is built in the shape and colors of a typical Swedish country house, to show the public that their homes are surrounded by radiation.

• www.next-up.org

This France-based organization brings together the latest research, regulations and European news on lobbying efforts and public action over EMR exposure from wireless communications.

• www.priartem.fr

PRIARTéM (Pour une réglementation des Implantations d'Antennes Relais de Téléphonie Mobile) is a citizens' group created in 2000 to monitor the installation of mobile phone antennas with respect to public health and safety. It was the first national organization in France to bring this issue to public view. With the rapid proliferation of mobile communications in France - 50 million cell phone users and nearly 100,000 antennas - the group's advocacy activities now include all wireless communications.

AUSTRALIA:

• www.emfacts.com

EMFacts was established in 1997 by Don Maisch as an independent source of information on the possible health and safety issues arising from human exposure to EMR. It is designed to be utilized as a resource by individuals, groups, organisations and communities who are trying to empower themselves by gaining a better understanding of the complex issues involved with this important environmental issue.

• www.emraustralia.com.au

The EMR Association of Australia is a consumer advocacy group which is working to alert people about the growing concerns in that country. It offers solutions to problems about electromagnetic radiation (EMR) for the workplace, local government and the home.

NEW ZEALAND:

• www.neilcherry.com

Dr. Neil Cherry (1946-2003) held the position of Associate Professor of Environmental Health at Lincoln University, New Zealand. Having listened to community concerns, Professor Cherry spent many years, and a great deal of his own funds, traveling around the world visiting universities and laboratories. He collected all the published papers and had personal discussions with the original researchers, to make sure the evidence and conclusions were sound and well-documented.

• www.notowers.co.nz

No Towers New Zealand's site presents general information and specifics about proposed cell towers and transmission line projects.

INTERNATIONAL:

• http://health.groups.yahoo.com/group/emfrefugee

EMF Refugee was created with the intent of bringing together electro-sensitive people in various countries to form their own EMF-free communities, where they can create healing environments for themselves and others, as well as the Earth.

• www.icems.eu

The International Commission for Electromagnetic Safety (ICEMS) is a non-profit organization registered in Italy that promotes research to protect public health from electromagnetic fields and develops the scientific basis and strategies for assessment, prevention, management and communication of risk, based on the precautionary principle.

Autism Related Issues

- www.generationrescue.org

Generation Rescue is an international movement of scientists and physicians researching the causes and treatments for autism, ADHD and chronic illness, while parent volunteers mentor thousands of families in recovering their children.

- website: www.thriiive.com

THRiiiVE!!! was founded by Dana Gorman. Their mission is to sort through the plethora of information available from all disciplines that impact health, organize it and disseminate it in a useful format so everyone can make informed choices that are beneficial to themselves and their family. She also brings together leading researchers and clinicians dealing with Autism Spectrum Disorders, and other neurological conditions.

Children & Nature

- www.childrenandnature.org

The Children & Nature Network (C&NN) was created to encourage and support the people and organizations working to reconnect children with nature. C&NN provides access to the latest news and research in the field and a peer-to-peer network of researchers and individuals, educators and organizations dedicated to children's health and wellbeing.

- www.childnature.ca

In North America there is a growing movement to reconnect children and families to nature for improved health and wellbeing. The Child and Nature Alliance is a non-profit organization in British Columbia that is formalizing the movement provincially, and eventually across Canada.

Medical Clinics/Professional Practices

Robert Abell, ND, LAc

Dr. Abell is a naturopathic physician who uses auricular medicine, homeopathy, acupuncture and Biotherapeutic Drainage as his primary therapeutic tools. He lectures on the assessment and treatments of neurological disorders including ADHD and autism. He offers a comprehensive understanding of the etiology and pathogenesis of a range of neurological disorders and methods of assessment, treatment and prevention in his educational programs. Dr. Abell treats children with ADHD and autism in his clinics and does phone consultations.

Be Well Acupuncture & Natural Medicine Clinic
Laguna Hills, California & Austin, Texas
Website: www.bewellhealthclinic.com

Leslie S. Feinberg, DC

Dr. Leslie S. Feinberg developed NeuroModulation Technique as a result of his many years of studying energetic medicine and the principles of western science and Traditional Chinese Medicine. Dr. Feinberg teaches this technique to other health professionals from many disciplines.

Columbia Chiropractic Clinic
NeuroModulation Technique NMT Seminars,
Hermiston, Oregon USA 97838
Website: www.nmt.md

Mark Hyman, MD and Elizabeth Boham, MD

The UltraWellness Center provides a functional medicine practice focusing on identifying and treating the underlying causes of illness, starting with extensive diagnostic testing. Dr. Mark Hyman is the author of several books on wellness issues including: environmental influences, supplementation, metabolism, detoxification and the conditions ADHD and autism. Dr. Hyman also offers training programs for health professionals through the Institute for Functional Medicine.

Lenox, MA
Email: office@ultrawellnesscenter.com
Website: www.drhyman.com

Carrie Hyman, LAc, OMD

Dr. Hyman is available for speaking engagements. Topics include: "Maintaining Health In The Wireless Age"; "Cell Phones, Kids & Us: The Inconvenient Truth About Convenient Technology" and "You Can't Fool Mother Nature: EMF & the Environment."

Contact: **dochyman1@gmail.com**

Dietrich Klinghardt, MD, PhD

Dr. Klinghardt was one of the first practitioners in the U.S. to include aggressive detoxification in his medical treatment strategies. Today many studies point in the same direction: chemical and heavy metal toxicity is at the core of many medical issues (cancer, neurological illnesses, fatigue, MS and others). Dr. Klinghardt has pioneered many diagnostic and detoxification strategies and is considered one of the worldwide leading authorities in this rapidly advancing field. There is an extensive list on the following website, under Referrals, of practitioners who have trained with Dr. Klinghardt.

Klinghardt Academy of Neurobiology
Website: **www.klinghardt.org**

Thomas Rau, MD

Paracelsus Klinik in Lustmühle Switzerland is the core of the international network of Paracelsus Biological Medicine. It was established in 1958 and is widely recognized as a center of excellence for natural medicine.

Websites: **www.drrau.com**

William J. Rea, MD

Dr. Rea is a thoracic and cardiovascular surgeon with a strong passion for the environmental aspects of health and disease. Dr. Rea is founder of The Environmental Health Center - Dallas, Texas, which offers comprehensive medical tests and treats human health problems including sensitivities to pollens, molds, dust, foods, chemicals, air (indoor/outdoor), water, electrical sensitivity (electromagnetic- EMF) and many more health problems as they relate to our environment.

Website: **www.ehcd.com**

Sanoviv Medical Institute

Dr. Myron Wentz founded this cutting edge medical and healing center, located south of San Diego, in Baja California. Sanoviv offers integrative, patient-centered medical treatments complemented by individualized personal care programs. Sanoviv guests receive functional medicine therapies to invigorate their immune system and treat many degenerative diseases, as well as counseling to help them continue to follow healthy lifestyles when they return to their homes.

Website: www.sanoviv.com

Hans-Christoph Scheiner, MD

Dr. Scheiner has been practicing holistic medicine in Munich, Germany, since 1975 with a focus on natural healing and psychotherapy. Dr. Scheiner is an author and presents medical expert opinions at parliamentary hearings. He is also a poet, dramatist, storyteller, composer and singer-songwriter. Dr. Scheiner's book, written with Ana Scheiner, is *Mobilfunk – die verkaufte Gesundheit*, 2006.

Dr. med. Hans-Christoph Scheiner
Institut für Holistische Medizin - München
Website: www.drscheiner-muenchen.de

Stephen T. Sinatra, MD, FACC, FACN, CNS,

Dr. Sinatra is a board-certified cardiologist, certified bioenergetic psychotherapist and certified nutrition and anti-aging specialist. He is a fellow in the American College of Cardiology and the American College of Nutrition, former chief of cardiology at Manchester Memorial Hospital, and an assistant clinical professor of medicine at the University of Connecticut School of Medicine. He writes a monthly national newsletter entitled Heart, Health and Nutrition (Healthy Directions, L.L.C.), lectures world-wide, and facilitates health-related workshops.

Website: www.heartmdinstitute.com

Jacob Teitelbaum, MD

Dr. Teitelbaum is a board certified internist and Medical Director of the national Fibromyalgia and Fatigue Centers, Inc., in the U.S. He focuses on expanding the understanding of hypothalamic dysfunction in Fibromyalgia (FMS) and Chronic Fatigue Syndrome (CFIDS). His work has led to the development of an effective treatment protocol for these diseases.

Website: www.endfatigue.com

Robert H. Weiner, PhD

Dr. Weiner is a Clinical Psychologist in Dallas, Texas who specializes in health psychology and behavioral medicine. He has just completed an international study that demonstrates the effectiveness of NeuroModulation Technique in treating autism. Further information about Dr. Weiner's practice and the NMT autism Study may be found online.

Website: www.living-solutions.com

To Find a Practitioner of Holistic/Complementary Medicine

Website: www.holisticmedicine.org

Biological Dentists

International Association of Mercury Free Dentists
Tel: (800) 335-7755
Website: www.dentalwellness4u.com

International Association of Biological Dentistry and Medicine
Tel: (281) 651-1745
Website: www.iabdm.org

Holistic Dental Association
Tel: (619) 923-3120
Website: www.holisticdental.org

Professional Services

Healthy Buildings:

International Institute for Bau-Biologie (IBE)
Website: www.buildingbiology.net

IBE/IBN Certified Consultants:

CANADA

Morriston, Ontario
Safe Living Technologies Inc.
Rob Metzinger, President
Testing equipment and protective/solution products, including shielding materials and air tube headsets
Email: metzinger@safelivingtechnologies.ca
Website: www.safelivingtechnologies.ca

Montreal, Quebec
3E Electro-magnetic Environmental Expertise Inc.
Stéphane Belainsky
Website: www.em3e.com

Victoria, British Columbia
Rainbow Consulting
Katharina Gustavs, Building Biology Environmental Consultant (IBN)
Website: www.buildingbiology.ca

USA

Carlsbad, California
Environmental Testing & Technology, Inc.(ET&T)
ET&T was founded in 1986 by Peter Sierck to provide an indoor environmental testing and consulting service for patients, physicians, home owners, corporations, institutions and builders. ET&T's EMF surveys have ranged from residential buildings, schools, computer interference investigations and mitigation, commercial buildings, work with local utilities on magnetic field reduction methods, to EMF studies for developers and EMF Management Plans for school districts.
Website: www.emfrf.com

San Diego, California
Gust Environmental
Larry Gust
Email: lgust@wavecable.com
Website: **www.healbuildings.com**

Clearwater, Florida
Indoor Environmental Technologies, Inc.
William H. Spates, III – BBEI
Website: **www.ietbuildinghealth.com**

Lyles, Tennessee
Wings of Eagles Healthy Living, Inc.
Vicki Warren, BBEC
Website: **www.wehliving.org**

Baltimore, Maryland
Healthy Home Arts, LLC
Jennifer Michalack, BBEC
Website: **www.healthyhomearts.com**

Battle Creek, Michigan
Mike Weston
Email: **westonmf@comcast.net**

St. Paul, Minnesota
Damon Coyne
E-Mail:damon@intentionalenvironment.com
Website: **intentionalenvironment.com**

Other EMF/EMR Consultants

CANADA

South Western British Columbia
ElectroSmogSolutions
Chris Anderson & Associates,
An EMR consultant and testing technician who has been following the Building Biology principles since 1999.
Website: **www.ElectroSmogSolutions.ca**

USA

Boulder, Colorado
Radsafe
Stan Hartman
Environmental health testing specializing in electro-magnetic fields and radiation.
Email: stan@radsafe.net
Website: **www.radsafe.net**

UK / NATIONWIDE

Offices in: Surrey, Yorkshire & Powys
Electric Forester Investigations Ltd.
Roger Moller BSc - Principal Investigator
On-Site and Remote Electromagnetic Surveys & Investigations
Homes, offices, factories and other built environments
Email: surveys@electricforester.co.uk
Website: **www.electricforester.co.uk**

EUROPE

Esslingen, Switzerland
Peter Schlegel
EMF/EMR testing of buildings; consulting in shielding; publicity & presentations in the field of EMF/EMR protection
Email: info@buergerwelle-schweiz.org
Website: **www.buergerwelle-schweiz.org**

AUSTRALIA

EMF/EMR Testing:

Tasmania/Melbourne:
EMFacts Consultancy
Don Maisch
Email: dmaisch@emfacts.com
Website: **www.emfacts.com**

Sydney:
EMR Surveys
John Lincoln
Email: emrsurveys@optusnet.com.au
Website: **www.emrsurveys.com.au**

Sydney:
EMR Australia
Lyn McLean
Email: contact@emraustralia.com.au
Website: **www.emrandhealth.com.au**

Melbourne, Victoria
Magshield Products, Intl.
Garry Melik
Email: **gmelik@magshield.com.au**

Melbourne, Victoria (Australia-wide)
Brookes EMS Pty Ltd.
Ken Brookes
Email: ken@brookes.com.au
Website: **www.brookesems.com.au**

Testing and Shielding Equipment

First, here is a brief description of each item; followed by sources listed by country.

Frequency Detection:

RF meters detect radio and microwave frequencies from mobile phones, wireless devices etc.

Electric Field meters detect lower frequencies from plugged-in devices, wiring etc.

Protective Shielding:

Air Tube, RF-Reducing Cell Phone Headsets – The sound travels up the plastic air tube, limiting the amount of radiation to your brain from either cell phones or portable phones. If you must use a cell phone, using this headset will reduce radiation to the brain significantly.

Bed Canopies – People who are electrically sensitive find shielding bed canopies essential, as do high performance athletes and, increasingly, business executives bathed in frequencies all day. Must be done properly; the material cannot rest on you and/or the bed.

Shielding Fabrics – Shielding fabric by the foot can be used to line curtains, walls and can even be put under carpets or desk chair pads to shield from the wireless radiation of neighbors downstairs.

Shielding Films – Placing plastic shielding film on glass windows can significantly block out ambient radio frequency radiation-from where it comes into the building most easily.

Shielding Paints – Metallic water-based paint can be placed under latex paints or wallpapers. Suitable for many types of exterior surfaces as well. Caution: The surface of this paint is highly conductive. It does need to be grounded and should not be installed without the services of an electrician.

Also, do not put shielding paint on the walls if the room will contain RF emitting devices, such as computers, cell phones, wireless routers, compact fluorescent bulbs, etc.

CANADA

Safe Living Technologies
Website: www.safelivingtechnologies.ca

RF Meters:
RF Detector – The Electrosmog Detector
This broadband detector allows you to hear and assess all the invisible microwave emissions around you.

Electric Field Meters:
eME3030B Dual Function Gauss Meter/Tesla Meter and Electric Field Detector

Shielding Materials – A full line of bed canopies, fabrics, window foil and paint.

Air tube, RF-Reducing Cell Phone Headsets – Avoid the health hazards of traditional headsets. This headset allows for safer cell phone use.

Demand Switches – These products eliminate Electric and Magnetic fields in living spaces by automatically shutting off the power when it is not being used.

Shielded Power Cord – Help reduce electric fields by replacing your existing power cords with shielded power cords.

THE USA

EMF Safety Store

Website: www.emfsafetystore.com

RF & Other Meters:

ELECTROSMOG METER (RF Fields)
Very Sensitive, Affordable RF/MW Meter

ZAP CHECKER (RF Fields)
Economical, Fast, Extremely Sensitive RF/MW

4180 ELF GAUSS METER (Magnetic Fields)

BUDGET BUZZ STICK (Magnetic Fields)
This little unit won't actually measure anything numerically, but is a great sniffer-out of magnetic fields, letting you listen to them.

ELFLX DETECTOR (Electric Fields)

AC Electric Field Finder

This inexpensive little unit has no numerical readout, but will detect AC electric fields from 100 to 600 volts (50-60 Hz), displaying them with a red LED and an audio output.

Air tube RF-Reducing Cell Phone Headsets.

UK/EUROPE:

Sensory Perspective

Website: **www.detect-protect.com**

RF Meters

The Electrosmog Detector — This broadband detector allows you to hear all the invisible Radio Frequency/microwave emissions around you, especially the hot spots.

HF32D — Digital Microwave Analyser for 800-2500 MHz - For a quick and competent digital assessment of microwave exposure between 800 MHz and 2.5 GHz. This range includes mobile/cellular phones, cordless phones, WiMAX, Bluetooth, and computer/broadband wLANs.

Electric Field Meters

ELF-3030B - Digital ELF Meter for E & H Fields 16Hz-2kHz

Sensitive and informative, with digital display and sounder, it allows you to check electromagnetic radiation from the delivery and use of electricity around you.

Shielding Materials — a full line of bed canopies, fabrics, wallpaper and paint, clothing and demand switches.

Publications

The Powerwatch Handbook — Simple Ways to Make Your (EMF) Environment Safer **(this is highly recommended.)**

SWEDEN:

TCO Development – Technology for You and the Planet

Website: www.tcodevelopment.com

This organization also has offices in Norway, Germany, the U.S. and Taiwan. It has developed a testing and certification process to ensure that information technology and office equipment meet standards of usability, while minimizing environmental impact.

Infrared Saunas

Heavenly Heat Saunas

Website: www.heavenlyheatsaunas.com

These far-infrared saunas are tested to assure lower electro-magnetic fields and have much lower fields than other infrared saunas. They are also made with the chemically sensitive person in mind.

Books

Balch, James F., MD and Balch, Phyllis A., CNC, *Prescription for Nutritional Healing* (Avery 4th edition, 2006)

Becker, Robert O., MD, and Selden, Gary, *The Body Electric – Electromagnetism and the Foundation of Life* (Harper, 1985)

Becker, Robert O., MD, *Cross Currents: The Perils of Electropollution, the Promise of Electromedicine* (Tarcher/Penguin, 1990)

Davis, Devra, PhD, *The Secret History of the War on Cancer* (Basic Books; Paperback edition, 2009) Website: www.environmentalheathtrust.org

Davis, Devra, PhD, *Sell Phone: What's Really On the Line* (Dutton Publishing, 2010) Website: www.environmentalheathtrust.org

Dean, Carolyn, MD, ND, *The Magnesium Miracle* (Ballentine, 2007) and *Complementary Natural Prescriptions for Common Ailments* (Keats Publishing Inc, 1994)

Goldberg, Gerald, MD, *Would You Put Your Head in a Microwave Oven? Microwave Radiation: An Emerging Healthcare Crisis* (AuthorHouse, 2006)

Grant, Lucinda, *The Electrical Sensitivity Handbook: How Electromagnetic Fields are Making People Sick* (Weldon Pub, 1995)

Hobbs, Angela, *The Sick House Survival Guide – Simple Steps to Healthier Homes* (New Society Publishers, 2003)

Hyman, Mark, MD, *The UltraMind Solution* (Simon & Schuster, 2009)

Kabat-Zinn, Jon, *Wherever You Go There You Are – Mindfulness Meditation in Everyday Life* (Hyperion 1st ed, 1994)

Levitt, B. Blake, editor, *Cell Towers – Wireless Convenience? Or Environmental Hazard?* (New Century Publishing, 2000)

Louv, Richard, *Last Child in the Woods – Saving Our Children From Nature Deficit Disorder* (Algonquin Books; Revised and annotated. ed, 2006)

McCarthy, Jenny & Kartzinel, Jerry, MD, *Healing and Preventing Autism* (Dutton Publishing, 2009)

McCarthy, Jenny, *Mother Warriors: A Nation of Parents Healing Autism Against All Odds* (Dutton Publishing, 2008)

Rea, William, MD, *Chemical Sensitivity*, 4 volumes (CRC Press and Lewis Publishers) from his website www.ehcd.com

Rea, William, MD, *Designing and Building a Healthy Home or Office: Optimum Environments for Optimum Health & Creativity* (2002)

Rees, Camilla and Havas, Magda *Public Health SOS: The Shadow Side of the Wireless Revolution* – available at www.electromagnetichealth.org

Scheiner, Hans-Christoph, Dr. med. & Scheiner, Ana, *Mobilfunk die verkaufte Gesundheit* (Michaels Verlag, 2006)

Sinatra, Stephen, MD, FACC, CNS, *Reverse Heart Disease Now* (John Wiley Publishing, 2007)

Sinatra, Stephen, MD, FACC, CNS, *The Sinatra Solution, Metabolic Cardiology* (Basic Health Publications, 2008)

Sinatra, Stephen, MD, FACC, CNS, *Earthing* (Basic Health Publications, 2010)

Teitelbaum, Jacob, MD, *From Fatigued to Fantastic* (Avery, a Member of Penguin/Putnam Ltd, 2001)

Vasey, Dr. Christopher, ND, *The Acid-Alkaline Diet for Optimum Health* (Healing Arts Press, 2nd edition, 2006)

Weil, Andrew, MD, *Healthy Aging: A Lifelong Guide to Your Well-Being* (Knopf, 2005)

Young, Robert O., PhD, DSc with Young, Shelley Redford, LMT, *Sick and Tired? Reclaim Your Inner Terrain* (Woodland Publishing, 2001)

Zhu, Hong Zhen, Dr. TCM, *Building a Jade Screen: Better Health with Chinese Medicine* (Penguin Canada, 2001)

Reference Material

The BioInitiative Report

Read the full report www.bioinitiative.org/report/index.htm

Cellphones and Brain Tumors - 15 Reasons for Concern

August 2009, Primary Author: L. Lloyd Morgan, USA, Bioelectromagnetics Society, Electronics Engineer (retired)

Summary

1: Industry's own research showed cell phones caused brain tumors.

2: Subsequent industry-funded research also showed that using a cell phone elevated the risk of brain tumors (2000-2002).

3: Interphone studies, published to date, consistently show use of a cell phone for less than 10 years protects the user from a brain tumor.

4: Independent research shows there is risk of brain tumors from cell phone use.

5: Despite the systemic-protective-skewing of all results in the Interphone studies, significant risk for brain tumors from cell phone use was still found.

6: Studies independent of industry funding show what would be expected if wireless phones cause brain tumors.

7: The danger of brain tumors from cell phone use is highest in children, and the younger a child is when he/she starts using a cell phone, the higher the risk.

8: There have been numerous governmental warnings about children's use of cell phones.

9: Exposure limits for cell phones are based only on the danger from heating.

10: An overwhelming majority of the European Parliament has voted for a set of changes based on 'health concerns associated with electromagnetic fields.'

11: Cell phone radiation damages DNA, an undisputed cause of cancer.

12: Cell phone radiation has been shown to cause the blood-brain barrier to leak.

13: Cell phone user manuals warn customers to keep the cell phone away from the body even when the cell phone is not in use.

14: Federal Communications Commission (FCC) has issued a warning for cordless phones.

15: Male fertility is damaged by cell phone radiation.

You can read the entire report at www.radiationresearch.org

From Larry Gust Environmental

Website: **www.healbuildings.com**

SYMPTOMS OF ELECTRO-MAGNETIC HYPERSENSITIVITY

Neurological:

Headaches, dizziness, nausea, difficulty concentrating, memory loss, irritability, depression, anxiety, insomnia, fatigue, weakness, tremors, muscle spasms, numbness, tingling, altered reflexes, muscle and joint paint, leg/foot pain, flu-like symptoms, fever.

More severe reactions can include seizures, paralysis, psychosis and stroke.

Cardiac:

Palpitations, arrhythmias, pain or pressure in the chest, low or high blood pressure, slow or fast heart rate, shortness of breath.

Respiratory:

Sinusitis, bronchitis, pneumonia, asthma.

Dermatological:

Skin rash, itching, burning, facial flushing.

Ophthalmologic:

Pain or burning in the eyes, pressure in/behind the eyes, deteriorating vision, floaters, cataracts.

Others:

Digestive problems, abdominal pain, enlarged thyroid, testicular/ovarian pain, dryness of lips, tongue, mouth, eyes, great thirst, dehydration, nosebleeds, internal bleeding, altered sugar metabolism, immune abnormalities, redistribution of metals within the body, hair loss, pain in the teeth, deteriorating fillings, impaired sense of smell, ringing in the ears.

Long Term Effects:

Adult cancer, tumors, childhood leukemia, breast cancer, DNA strand breakage, abnormal cell division, nerve damage, MS, ALS, Alzheimer and Parkinson disease, brain damage, melatonin reduction, miscarriages.

Biological Mechanisms:

Some suggest that these epidemiological studies should be rejected because they claim that there are no known biological mechanisms. This is wrong on two counts. First, epidemiological evidence is the strongest evidence of human health effects, and dose-response relationships are indicative of a causal effect, Hill (1965).

Biological mechanisms are limited by current knowledge and therefore should not diminish the epidemiological conclusions. Second, there is a large and coherent body of evidence of biological mechanisms that support the conclusion of a plausible, logical and causal relationship between EMR exposure and cancer, cardiac, neurological and reproductive health effects.

Neurological Interactions:

König (1974) and Wever (1974) prove that ELF EMR interacts with and interferes with human brains at extremely low field intensities.

Calcium Ion Homeostasis:

Blackman (1990) concludes that there is strong evidence that EMR alters cellular calcium ion homeostasis, down to 0.08 mW/cm^2, Schwartz et al. (1990).

Chromosome Aberrations:

Fourteen studies show that RF/MW causes significant chromosome damage, four with dose response relationships and one recorded a dose related cell death rate; Heller and Teixeira-Pinto (1959), Tonascia and Tonascia (1996) [cited in Goldsmith (1997b)], Sagripanti and Swicord (1986), Garaj-Vrhovac et al. (1990, 1991, 1992, 1993, 1998), Maes et al. (1993), Timchenko and Ianchevskaia (1995), Balode (1996), Haider et al. (1994), Vijayalaxmi et al. (1997), Tice, Hook and McRee (1999).

DNA strand breakage:

Four independent laboratories observe significant DNA damage, including two for cell phone radiation, down to 1 mW/cm^2, Phillips et al. (1998). Lai and Singh (1995, 1996, 1997), Sarkar, Ali and Behari (1994), Verschave et al. (1994), including a dose response relationship, Lai and Singh (1996).

Neoplastic Transformation of Cells:

Balcer-Kubiczek and Harrison (1991) observed a significant dose response in cells exposed to microwaves.

Oncogene Activity:

Two laboratories show that cell phone radiation significantly alters proto oncogene activity; Ivaschuk et al. (1997) and Goswami et al. (1999).

Melatonin Reduction:

Fourteen studies show that EMR across the spectrum from ELF to RF/MW reduces melatonin in people.

Wang (1989) who found that workers who were more highly exposed to RF/MW had a dose-response increase in serotonin, and hence indicates a reduction in melatonin. Abelin (1999) reported significant reductions from SW radio exposure, Burch et al. (1997) with a combination of 60 Hz fields and cell phone use and Arnetz et al. (1996) with VDTs.

ELF exposure reduced melatonin in Wilson et al. (1990), Graham et al. (1994), Wood et al. (1998), Karasek et al. (1998), and Burch et al. (1997, 1998, 1999a), Juutilainen et al. (2000) and Graham et al. (2000); Pfluger et al. (1996)[16.7 Hz] and geomagnetic activity, Burch et al. (1999b).

Immune system impairment by EMR

Impairment of the immune system is related to calcium ion efflux, Walleczek (1992) and to reduced melatonin, Reiter and Robinson (1995). Cossarizza et al. (1993) showed that ELF fields increased both the spontaneous and PHA and TPA- induced production of interleukin-1 and IL-6 in human peripheral blood.

Rats exposed to microwaves showed a significant reduction in splenic activity of natural killer (NK) cells, Nakamura et al. (1997).

Quan et al. (1992) showed that microwave heating of human breast milk highly significantly suppressed the specific immune system factors for E. Coli bacteria compared with conventional heating. Dmoch and Moszczynski (1998) found that microwave exposed workers had decreased NK cells and a lower value of the T-helper/T-suppressor ratio was found. Moszczynski et al. (1999) observed increased IgG and IgA and decreased lymphocytes and T8 cells in TV signal exposed workers.

Chronic, 25 year, exposure to an extremely low intensity (<0.1 mW/cm^2) 156-162 MHz, 24.4 Hz pulse frequency, radar signal in Latvia produced significant alterations in the immune system factors of exposed villagers, Bruvere et al. (1998).

Biological Mechanism Conclusions:

EMR is shown to alter cellular calcium ions, significantly increase chromosome aberrations, DNA strand breakage, neoplastic transformation of cells, reduce melatonin, enhance oncogene activity and impair the immune system.

This is a coherent, consistent and overwhelming set of evidence to show that EMR is genotoxic.

When coupled with the epidemiological evidence of cancer, there is compelling evidence that EMR is genotoxic, and, hence, is carcinogenic and teratogenic.

SCIENTISTS' CONCERNS ABOUT WIRELESS IN SCHOOLS

Open Letter to Parents, Teachers, School Boards. Regarding Wi-Fi Networks in Schools - May 5, 2009

by Dr. Magda Havas, Associate Professor, Trent University

FACTS:
1. GUIDELINES

Guidelines for microwave radiation (which is what is used in Wi-Fi) range 5 orders of magnitude in countries around the world. The lowest guidelines are in Salzburg, Austria, and now in Liechtenstein. The guideline in these countries is 0.1 microW/cm^2. In Canada it is 1,000 microW/cm^2! Why does Canada have guidelines that are so much higher than other countries?

Canada's guidelines are based on a short-term (6-minute) heating effect. It is assumed that if this radiation does not heat your tissue it is "safe." This is not correct. Effects are documented at levels well below those that are able to heat body tissue (Analysis of Health and Environmental Effects of Proposed San Francisco Earthlink Wi-Fi Network, 2007). These biological effects include increased permeability of the blood-brain barrier, increased calcium flux, increase in cancer and DNA breaks, induced stress proteins, and nerve damage. Exposure to this energy is associated with altered white blood cells in school children; childhood leukemia; impaired motor function, reaction time, and memory; headaches, dizziness, fatigue, weakness, and insomnia.

2. ELECTRO-HYPER-SENSITIVITY

A growing population is adversely affected by these electromagnetic frequencies. The illness is referred to as "electro-hyper-sensitivity" (EHS) and is recognized as a disability in Sweden. The World Health Organization defines EHS as: ". . . a phenomenon where individuals experience adverse health effects while using or being in the vicinity of devices emanating electric, magnetic, or electromagnetic fields (EMFs). . . EHS is a real and sometimes a debilitating problem for the affected persons, while the level of EMF in their neighborhood is no greater than is encountered in normal living environments. Their exposures are generally several orders of magnitude under the limits in internationally accepted standards."

Health Canada acknowledges in their Safety Code 6 guideline that some people are more sensitive to this form of energy, but they have yet to address this by revising their guidelines. Symptoms of EHS include sleep disturbance, fatigue, pain, nausea, skin disorders, problems with eyes and ears (tinnitus), dizziness, etc. It is estimated that 3% of the population are severely affected and another 35% have moderate symptoms. Prolonged exposure may be related to sensitivity and for this reason it is imperative that children's exposure to microwave radiation (Wi-Fi and mobile phones) be minimized as much as possible.

3. CHILDREN'S SENSITIVITY

Children are more sensitive to environmental contaminants and that includes microwave radiation. The Stewart Report (2000) recommended that children not use cell phones except for emergencies. The cell phone exposes your head to microwave radiation. A wireless computer (Wi-Fi) exposes your entire upper body and if you have the computer on your lap it exposes your reproductive organs as well. Certainly this is not desirable, especially for younger children and teenagers. For this reason we need to discourage the use of wireless technology by children, especially in elementary schools. That does not mean that students cannot go on the Internet. It simply means that access to the Internet needs to be through wires rather than through the air (wireless, Wi-Fi).

4. REMOVAL OF WI-FI

Most people do not want to live near either cell phone antennas or Wi-Fi antennas because of health concerns. Yet when Wi-Fi (wireless routers) are used inside buildings it is similar to the antenna being inside the building rather than outside and is potentially much worse with respect to exposure since you are closer to the source of emission.

Libraries in France are removing Wi-Fi because of concern from both the scientific community and their employees and patrons. The Vancouver School Board (VSB) passed a resolution in January 2005 that prohibits construction of cellular antennas within 1000 feet (305 m) of school property. Palm Beach, Florida, Los Angeles, California, and New Zealand have all prohibited cell phone base stations and antennas near schools due to safety concerns. The decision not to place cell antennas near schools is based on the likelihood that children are more susceptible to this form of radiation. Clearly if we do not want antennas "near" schools, we certainly do not want antennas "inside" schools! The safest route is to have wired Internet access rather than wireless. While this is the more costly alternative in the short-term it is the least costly alternative in the long run if we factor in the cost of ill health of both teachers and students.

5. ADVISORIES

Advisories to limit cell phone use have been issued by various countries and organizations including the UK (2000), Germany (2007), France, Russia, India, Belgium (2008) as well as the Toronto Board of Health (July 2008) and the Pittsburgh Cancer Institute (July 2008). While these advisories relate to cell phone use, they apply to Wi-Fi exposure as well since both use microwave radiation. If anything, Wi-Fi computers expose more of the body to this radiation than do cell phones.

6. PRECAUTIONARY PRINCIPLE

Even those who do not "accept" the science showing adverse biological effects of microwave exposure should recognize the need to be careful with the health of children. For this reason, we have the Precautionary Principle, which states: In order to protect the environment, the precautionary approach shall be widely applied by States according to their capability. Where there are threats of serious or irreversible damage, lack of full scientific certainty shall not be used as a reason for postponing cost effective measures to prevent environmental degradation. In this case "States" refers to the School Board and those who make decisions about the health of children.

The two most important environments in a child's life are the home (especially the bedroom) and the school. For this reason it is imperative that these environments remain as safe as possible. If we are to err, please let us err on the side of caution.

Columbia University, College of Physicians and Surgeons
Department of Physiology and Cellular Biophysics

Re: Health effects of cell tower radiation

As an active researcher on biological effects of electromagnetic fields (EMF) for over twenty five years at Columbia University, as well as one of the organizers of the 2007 online Bioinitiative Report on the subject, I am writing in support of a limit on the construction of cell towers in the vicinity of schools.

There is now sufficient scientific data about the biological effects of EMF, and in particular about radiofrequency (RF) radiation, to argue for adoption of precautionary measures. We can state unequivocally that EMF can cause single and double strand DNA breakage at exposure levels that are considered safe under the FCC guidelines in the USA. As I shall illustrate below, there are also epidemiology studies that show an increased risk of cancers associated with exposure to RF. Since we know that an accumulation of changes or mutations in DNA is associated with cancer, there is good reason to believe that the elevated rates of cancers among persons living near RF towers are probably linked to DNA damage caused by EMF. Because of the nature of EMF exposure and the length of time it takes for most cancers to develop, one cannot expect 'conclusive proof' such as the link between helicobacter pylori and gastric ulcer. (That link was recently demonstrated by the Australian doctor who proved a link conclusively by swallowing the bacteria and getting the disease.) However, there is enough evidence of a plausible mechanism to link EMF exposure to increased risk of cancer, and therefore of a need to limit exposure, especially of children.

EMF have been shown to cause other potentially harmful biological effects, such as leakage of the blood-brain barrier that can lead to damage of neurons in the brain, increased micronuclei (DNA fragments) in human blood lymphocytes, all at EMF exposures well below the limits in the current FCC guidelines. Probably the most convincing evidence of potential harm comes from living cells themselves when they start to manufacture stress proteins upon exposure to EMF. The stress response occurs with a number of potentially harmful environmental factors, such as elevated temperature, changes in pH, toxic metals, etc. This means that when stress protein synthesis is stimulated by radiofrequency or power frequency EMF, the body is telling us in its own language that RF exposure is potentially harmful.

There have been several attempts to measure the health risks associated with exposure to RF, and I can best summarize the findings with a graph from the study by Dr. Neil Cherry of all childhood cancers around the Sutro Tower in San Francisco between the years 1937 and 1988. Similar studies with similar results were done around broadcasting antennas in Sydney, Australia and Rome, Italy, and there are now studies of effects of cellphones on brain cancer. The Sutro tower contains antennas for broadcasting FM (54.7 kW), TV (616 kW) and UHF (18.3 MW) signals over a fairly wide area, and while the fields are not uniform, and also vary during the day, the fields were measured and average values estimated, so that one could associate the cancer risk with the degree of EMF exposure.

The data in the figure are the risk ratios (RR) for a total of 123 cases of childhood cancer from a population of 50,686 children, and include 51 cases of leukaemia, 35 cases of brain cancer and 37 cases of lymphatic cancer. It is clear from the results that the risk ratio for all childhood cancers is elevated in the area studied, and while the risk falls off with radial distance from the antennas, as expected, it is still above a risk ratio of 5 even at a distance of 3 km where the field was $1\mu W/cm^2$. This figure is what we can expect from prolonged RF exposure. In the Bioinitiative Report, we recommended $0.1\mu W/cm^2$ as a desirable precautionary level based on this and related studies, including recent studies of brain cancer and cellphone exposure.

As I mentioned above, many potentially harmful effects, such as the stress response and DNA strand breaks, occur at nonthermal levels (field strengths that do not cause a temperature increase) and are therefore considered safe. It is obvious that the safety standards must be revised downward to take into account the nonthermal as well as thermal biological responses that occur at much lower intensities. Since we cannot rely on the current standards, it is best to act according to the precautionary principle, the approach advocated by the European Union and the scientists involved in the Bioinitiative report. In light of the current evidence, the precautionary approach appears to be the most reasonable for those who must protect the health and welfare of the public and especially its most vulnerable members, children of school-age.

Sincerely yours,
Martin Blank, PhD
Associate Professor of Physiology and Cellular Biophysics

KEY STUDIES CITED BY DR. SARAH STARKEY

Website: www.wifiinschools.org.uk

Aitken R. J., Bennetts L.E., Sawyer D., Wikiendt A. M. and King B. V., 2005, Impact of radio frequency electromagnetic radiation on DNA integrity in the male germline, Int J Androl, 28(3), 171-179.

Arendt J., Labib M. H., Bojkowski C., Hanson S. and Marks V., 1989, Rapid decrease in melatonin production during successful treatment of delayed puberty, Lancet 1(8650), 1326.

Bio-Initiative Report, 2007, A Rationale for a biologically-based public exposure standard for electromagnetic fields (ELF and RF), http://www.bioinitiative.org/index.htm (accessed August 2008).

Burch J. B., Reif J. S., Noonan C. W., Ichinose T., Bachand A. M., Koleber T. L. and Yost M. G., 2002, Melatonin metabolite excretion among cellular telephone users, International Journal of Radiation Biology 78(11), 1029-1036.

Divan H. A., Kheifets L., Obel C. and Olsen J., 2008, Prenatal and postnatal exposure to cell phone use and behavioural problems in children, Epidemiology 19(4), 523-529.

Johansson, O., 2006, Electrohypersensitivity: Observations in the human skin of a physical impairment, p.107-117, Proceedings of the International Workshop on EMF Hypersensitivity, Prague, 2004

Erogul O., Oztas E., Yildirim I., Kir T., Aydur E., Komesli G., Irkilata H. C., Irmak M. K. and Peker A. F., 2006, Effects of electromagnetic radiation from a cellular phone on human sperm motility: an in vitro study, Arch Med Res 37(7), 840-843.

Huber R., Schuderer J., Graf T., Jatz K., Borbaly A. A., Kuster N. and Achermann P., 2003, Radio frequency electromagnetic field exposure in humans: Estimation of SAR distribution in the brain, effects on sleep and heart rate, Bioelectromagnetics 24(4), 262-267.

Lai, H., 2007, Evidence for effects on neurology and behaviour, Bio-Initiative Report, www.bioinitiative.org (accessed August '08).

Landgrebe M., Hauser S., Langguth B., Frick U., Hajak G. and Eichhammer P., 2007, Altered cortical excitability in subjectively electrosensitive patients: results of a pilot study, J. Psychosom Res., 62(3), 283-288.

Lopez-Martin E., Relova-Quinteiro J. L., Gallego-Gomez R., Peleteiro-Fernandez M., Jorge-Barreiro F. J. and Ares-Pena F. J., 2006, GSM radiation triggers seizures and increases cerebral c-fos positivity in rats pretreated with subconvulsive doses of picrotoxin, Neuroscience Letters 398, 139-144.

Maachi M. M., and Bruce J. N., 2004, Human pineal physiology and functional significance of melatonin, Frontiers in Neuroendocrinology 25, 177-195.

Maby E., Le Bouquin R. and Faucon G., 2006, Short-term effects of GSM mobile phones on spectral components of the human electroencephalogram, Proceedings of the 28th IEEE EMBS Annual International Conference 1, 3751-3754.

Maier R., Greter S.-E. and Maier N., 2004, Effects of pulsed electromagnetic fields on cognitive processes - a pilot study on pulsed field interference with cognitive regeneration, Acta Neurol Scand 110, 46-52.

Murcia Garcia J., Munoz Hoyos A., Molina Carballo A., Fernandez Garcia J. M., Narbona Lopezet E. and Fernandez J. U., 2002, Puberty and melatonin, An Esp Pediatr 57(2), 121-126.

Nittby H., Grafstram G., Tian D. P., Malmgren L., Brun A., Persson B. R., Salford L. G., Eberhardt J., 2008, Cognitive impairment in rats after long-term exposure to GSM-900 mobile phone radiation, Bioelectromagnetics 29(3), 219-232.

Pyrpasopoulou A., Kotoula V., Cheva A., Hytiroglou P., Nikolakaki E., Magras I. N., Xenos T.D., Tsiboukis T. D. and Karkavelas G., 2004, Bone morphogenetic protein expression in newborn rat kidneys after prenatal exposure to radiofrequency radiation, Bioelectromagnetics 25(3), 216-227.

Stewart Report, 2000, Independent Expert Group on Mobile Phones, http://www.iegmp.org.uk/report/index.htm (accessed August '08).

Vecchio F., Babiloni C., Ferreri F., Curcio G., Fini R., Del Percio C. and Rossini P. M., 2007, Mobile phone emission modulates interhemispheric functional coupling of EEG alpha rhythms, European Journal of Neuroscience 25, 1908-1913.

Waldhauser F., Boepple P. A., Schemper M., Mansfield M. J. and Crowley Jr W. F., 1981, Serum melatonin in central precocious puberty is lower than in age-matched prepubertal children, J. Clin Endocrimol. Metab. 73, 793-796.

Wdowiak A., Wdowiak L. and Wiktor H., 2007, Evaluation of the effect of using mobile phones on male fertility, Ann Agric Environ Med 14, 169-172.

Wiart J., Hadjem A., Wong M. F., Bloch I., 2008, Analysis of RF exposure in the head tissues of children and adults, Phys Med Biol 53(13), 3681-3695.

WHO (World Health Organisation), 2006, Electromagnetic Hypersensitivity, Proceedings of the International Workshop on EMF Hypersensitivity, Prague, 2004, https://www.who.int/peh-emf/publications/reports/EHS_Proceedings_June2006.pdf

WIRELESS IN THE AIR - HEALTH AND SAFETY CONCERNS

This is the full letter written to many airline executives by Dr. Scheiner and the German Environmental Physician Initiative.

Munich, July 24th, 2008

Intended Authorization of Mobile Phones and Wireless LAN in Air Traffic

Dear ─────────,

The information was spread in the media, that it is planned to allow the use of wireless communication devices like cell phones, W-LAN, and similar electronic tools on commercial flights. We are highly concerned about this fact.

As a technical innovation the susceptibility of electronic board systems in relation to microwaves has decreased (to be questioned here are the remaining risks): the personal use of wireless communication technology on commercial flights leads to serious health risks to all passengers and flight personnel. Therefore it should be treated as a fact of the overall security of commercial airlines.

Reason: If there are wireless systems permanently in use like cell-phone stations, W-LAN, Blue-Tooth, DECT, etc. in addition to active single cell phones and notebooks on flights with the duration of several hours, the passengers, crew and pilots would be exposed to excessive radiation of 25,000 nW/cm^2 and higher.

Even though these levels of exposure are just from 1/10 to 1/50 of the actual legal exposure limits, there are multiple scientific proofs of health risks. Even a radiation dose of 100-500 nW/cm^2 breaks the blood-brain barrier, which causes the entry of water, dissolving metabolism waste products, environmental toxins and blood proteins (especially albumins) into the central nervous system.

The fatal consequences are: miniature edemas occur in the complete brain, multiple selective swellings in non-renewable brain cells are irretrievably squeezed to death. They occur as dark neurons in the microscopic picture. Those dark neurons are proven to be possible starting points of very serious neurodegenerative diseases like Multiple Scleroses, Parkinson's disease, Alzheimer's disease, senile dementia and so on.

Many scientific studies with animal tests showed significantly this opening of the blood-brain barrier. Even low levels of 100-500 nW/cm² caused in more than 50% of the animals tested an opening of the blood-brain barrier. At the actual levels of 25,000 nW/cm², which are expected to occur on board of commercial airplanes, 100% of the animals tested had serious brain damage.

The breakage of the blood-brain barrier under the influence of radio waves and high frequency, which happens at levels far underneath the current legal exposure limit, has obtained doubtless scientific evidence. This was significantly found and described by ALBERTS 1977, OSCAR AND HAWKINS 1977, NEILLY AND LIN 1986 SALFORD, BRUN, PERSSON, 1994,1997,2003 AUBINEAU AND TOERE 2002, 2003 SCHIRMACHER 1999,2000 and many others.

Another fact: While the airplane is moving at an altitude between 8,000 and 12,000 metres, a reduced air pressure occurs inside of the plane, which equals the air pressure of 2000 – 3000 metres outside. Therefore the breakage of the blood-brain barrier is more likely because of the lack of oxygen and the well-known altitude sickness.

The severe consequences of the brain and nerve damage and safety of passengers are very concerning, especially those of the pilots, because they are already highly exposed from radar. The symptoms caused by high frequencies, like headaches, drowsiness, vertigo, nausea are often connected with loss of hearing and vision; lack of concentration and memorization are in this context known as the "Microwave-Syndrome" (JOHNSON-LIAKOURIS, 1998, MILD 1998, SANTINI 2001, 2002, 2003, NAVARRO, OBERFELD 2003).

Another dangerous result is the extreme slow down of the neuro-muscular response because of a doubled reaction time. Also the mental capacity is, in terms of cognitive disorders, verifiably heavily affected. Epidemiological studies and exposure trials with volunteers and animals show this clearly, see also the TNO-Study of PROFESSOR ZWAMBORN 2003, KALODYSKI U. KALODYNSKA 1996, PROFESSOR LAI U. SINGH 1966, 1997, 1998, ALTPETER U. ABELIN 1995, 1999, SEMM U. BEASOND 1996, RÖSCHKE UND MANN 1996 and many more.

Because of the mostly fatal exits of flight accidents there are no special data about the influence from radio- and microwaves on flight safety and security available. But the knowledge we have from other traffic systems on the ground is surely transferable: in 2002, The British Transport Research Laboratory found out that the time of reaction of a car driver is 30% lower if he has been exposed to radio waves than the reaction time of an alcoholized driver, and 50% lower than the reaction time of a driver who has not been exposed to either.

In 1997 the University of Toronto (REDELMEIER AND TIBSCHIRANI) found out on the basis of a big trial that in relation to the length of exposure to radio waves of the car drivers, the drivers were 5 times as likely to cause an accident and twice as likely to cause a deadly crash. The same was confirmed by the RESEARCH GROUP OF VIOLANTI (1998) "CELLULAR PHONES AND FATAL TRAFFIC COLLISIONS", similar facts were proved by PROFESSOR UNGER AT BREMEN UNIVERSITY: Cell Phone influence while driving leads even for experienced drivers to a 30% increase in changing and stopping mistakes!

In this context the following is very interesting: it has been proven multiple times that the use of cellular phones cause electroencephalography changes of the brain waves! Because of this it was possible for DR. VON KLITZING AT LUEBECK UNIVERSITY to prove that highly pulsed frequencies like mobile phones lead to pathologically EEG-patterns of the brain in the so called "alpha–rhythm". This EEG area represents our physical and mental relaxing and recovering phases. The "alpha-rhythm" is shown while we sleep and dream.

Pathological EEG changes in this alpha-area – show up most at 10 Hz – are an indication of a deep radio wave caused disorder of our physical and mental health, which reaches deep into our subconscious. Those EEG changes were already proven before DR. KLITZING by Russian and American researchers, and after him often reproduced, for example by the German Department for Work Safety and Occupational Medicine in 1998 (FREUDE ET. AL.), by THE UNIVERSITY OF ZUERICH UNDER PROF. ACHERMAN, HUBER, BORBELEY ET AL. 1999,2000,2002,2003).

The above explanations include that radio waves also cause serious sleep disorders. It is also proven multiple times that radio waves cause a decrease in the sleep and body defense hormone Melatonin. (BURCH U.A. 1997, 1998, 1999, 2000, REITER U. ROBINSON 1994,1995, ABELIN U. ALTPETER 1995, 1999 U.A.M.)

Let's not forget about the intermediate massive impairment of our microcirculation and therewith the oxygen supply of our inner organs and brain caused by the tendency of our red blood cells to stick together under the influence of radio waves. (DR. PETERSOHN 1998, RITTER UND WOLSKI 2005). In addition to this Prof. Kundi (Environmental hygienic department of Vienna University) found a high increase of heart attacks, strokes, thromboses and embolism in people who live near transmitter masts. All of these are disorders that could lead to an immediate airplane crash, with hundreds of victims, if this would happen to pilots.

And finally the following is for all airlines to consider: electrosensitivity and electroallergies to wireless electronic devices already bother 10% of the worldwide population, and this tendency is rapidly increasing. Even short flights, but especially long ones, on which the passengers are constantly given a continual exposure of radio and microwaves would lead to a reduction of bookings and a recognizable decrease in sales for airlines which permit cell phone systems, notebooks and similar wireless instruments on board. This has already been seen in Germany and other countries where electro-sensitive or allergic people avoid high-speed trains because of permanently active "Repeaters" (amplifiers of HF-signals). This is one of the facts that cause the above described microwave syndrome (headaches, nausea, drowsiness, vision disorders etc) and their change over to using different ways of transportation like cars, buses and trains without Repeaters.

The decrease of sales is doubtless continuing, and will do so even more as passengers realize what kind of danger permanent radio and microwaves on board hold for flight security.

We ask you not to ignore these scientific facts about the dangers of radio and microwaves.

With kind regards,
Dr. med. Hans-Christoph Scheiner
Environmental physician, Münchner Ärzteappell
By orders of the German Environmental Physician Initiative represented by:

Dr. med. Wolf Bergmann	Freiburger Ärzte-Appell
Dr. med. Horst Eger	Ärztlicher Qualitätszirkel Naila
Dr. med. Markus Kern	Mobilfunk-Ärzteinitiative Allgäu-Bodensee-Oberschwaben
Dr. med. Peter Lackner	Münchner Ärzteappell
Dr. med. dent. Joh. Lechner	München, Vorsitzender der GZM
Dr. med. Joachim Mutter	Universitätsklinik Freiburg

Summarizing the Science – Dr. Henry Lai

For those who argue there's not enough science to show any links between wireless communications and health effects, Dr. Henry Lai has kindly allowed me to share his detailed summary of studies conducted over the past couple of decades. Many of these show wireless communication signals affect cancers, immunology, genetics, cellular & molecular biology, behavior, reproduction and growth, hormones, the nervous system, the blood-brain barrier, metabolism, etc.

Reporting biological effects of radiofrequency radiation (RFR) at low intensities

(1) Balode (1996)- blood cells from cows from a farm close and in front of a radar showed significantly higher level of severe genetic damage.

(2) Boscol et al. (2001)- RFR from radio transmission stations (0.005 mW/cm^2) affects immunological system in women.

(3) Chiang et al. (1989)- people who lived and worked near radio antennae and radar installations showed deficits in psychological and short-term memory tests.

(4) de Pomerai et al. (2000, 2002) reported an increase in a molecular stress response in cells after exposure to a RFR at a SAR of 0.001 W/kg. This stress response is a basic biological process that is present in almost all animals - including humans.

(5) De Pomerai et al. (2003) RFR damages proteins at 0.015-0.02 W/kg.

(6) D'Inzeo et al. (1988)- very low intensity RFR (0.002–0.004 mW/cm^2) affects the operation of acetylcholinerelated ion-channels in cells. These channels play important roles in physiological and behavioral functions.

(7) Dolk et al. (1997)- a significant increase in adult leukemias was found in residents who lived near the Sutton Coldfield television (TV) and frequency modulation (FM) radio transmitter in England.

(8) Dutta et al. (1989) reported an increase in calcium efflux in cells after exposure to RFR at 0.005 W/kg. Calcium is an important component of normal cellular functions.

(9) Fesenko et al. (1999) reported a change in immunological functions in mice after exposure to RFR at a power density of 0.001 mW/cm^2.

(10) Hjollund et al. (1997)- sperm counts of Danish military personnel, who operated mobile ground-to-air missile units that use several RFR

emitting radar systems (maximal mean exposure 0.01 mW/cm²), were significantly lower compared to references.

(11) Hocking et al. (1996)- an association was found between increased childhood leukemia incidence and mortality and proximity to TV towers.

(12) Ivaschuk et al. (1999)- short-term exposure to cellular phone RFR of very low SAR (26 mW/kg) affected a gene related to cancer.

(13) Kolodynski and Kolodynska (1996)- school children who lived in front of a radio station had less developed memory and attention, their reaction time was slower, and their neuromuscular apparatus endurance was decreased.

(14) Kwee et al. (2001)- 20 minutes of cell phone RFR exposure at 0.0021 W/kg increased stress protein in human cells.

(15) Lebedeva et al. (2000)- brain wave activation was observed in human subjects exposed to cellular phone RFR at 0.06 mW/cm².

(16) Magras and Xenos (1999) reported a decrease in reproductive function in mice exposed to RFR at power densities of 0.000168 - 0.001053 mW/cm².

(17) Mann et al. (1998)- a transient increase in blood cortisol was observed in human subjects exposed to cellular phone RFR at 0.02 mW/cm². Cortisol is a hormone involved in stress reaction.

(18) Marinelli et al. (2004)- exposure to 900-MHz RFR at 0.0035 W/kg affected cell's self-defense responses.

(19) Michelozzi et al. (1998)- leukemia mortality within 3.5 km (5,863 inhabitants) near a high power radio transmitter in a peripheral area of Rome was higher than expected.

(20) Michelozzi et al. (2002)- childhood leukemia higher at a distance up to 6 km from a radio station.

(21) Navakatikian and Tomashevskaya (1994)- RFR at low intensities (0.01 - 0.1 mW/cm²; 0.0027- 0.027 W/kg) induced behavioral and endocrine changes in rats. Decreases in blood concentrations of testosterone and insulin were reported.

(22) Novoselova et al. (1999)- low intensity RFR (0.001 mW/cm²) affects functions of the immune system.

(23) Novoselova et al. (2004)- chronic exposure to RFR (0.001 mW/cm²) decreased tumor growth rate and enhanced survival in mice.

(24) Park et al. (2004)- higher mortality rates for all cancers and leukemia in

some age groups in the area near the AM radio broadcasting towers.

(25) Persson et al. (1997)- reported an increase in the permeability of the blood-brain barrier in mice exposed to RFR at 0.0004 - 0.008 W/kg. The blood-brain barrier envelops the brain and protects it from toxic substances.

(26) Phillips et al. (1998)- reported DNA damage in cells exposed to RFR at SAR of 0.0024 - 0.024 W/kg.

(27) Polonga-Moraru et al. (2002)- change in membrane of cells in the retina (eye) after exposure to RFR at 15 $\mu W/cm^2$.

(28) Pyrpasopoulou et al. (2004)- exposure to cell phone radiation during early gestation at SAR of 0.0005 W/kg (5 $\mu W/cm^2$) affected kidney development in rats.

(29) Salford et al. (2003)- nerve cell damage in brain of rats exposed for 2 hrs to GSM signal at 0.02 W/kg.

(30) Santini et al. (2002)- increase in complaint frequencies for tiredness, headache, sleep disturbance, discomfort, irritability, depression, loss of memory, dizziness, libido decrease, in people who lived within 300 m of mobile phone base stations.

(31) Sarimov et al. (2004)- GSM microwaves affect human lymphocyte chromatin similar to stress response at 0.0054 W/kg.

(32) Schwartz et al. (1990)- calcium movement in the heart affected by RFR at SAR of 0.00015 W/kg. Calcium is important in muscle contraction. Changes in calcium can affect heart functions.

(33) Somosy et al. (1991)- RFR at 0.024 W/kg caused molecular and structural changes in cells of mouse embryos.

(34) Stagg et al. (1997)- glioma cells exposed to cellular phone RFR at 0.0059 W/kg showed significant increases in thymidine incorporation, which may be an indication of an increase in cell division.

(35) Stark et al. (1997)- a two to seven-fold increase of salivary melatonin concentration was observed in dairy cattle exposed to RFR from a radio transmitter antenna.

(36) Tattersall et al. (2001)- low-intensity RFR (0.0016 - 0.0044 W/kg) can modulate the function of a part of the brain called the hippocampus, in the absence of gross thermal effects. The changes in excitability may be consistent with reported behavioral effects of RFR, since the hippocampus is involved in learning and memory.

Index

A

Abell, Robert (ND, LAc) 388
absorption of electro-magnetic radiation 71
AC magnetic fields 78
acupuncture 320
ADD 57, 115, 260, 286, 293
Adey, Dr. Ross 51, 52, 89
ADHD 4, 57, 115, 273, 283, 293, 295, 296, 298, 387, 388
ADSL modem/router 170
airplane travel 221
  in-flight video screens 223
air pollution in your home 270
airport 195
alarm clocks/clock radios 211
  Clocks - Safer Solutions 211
allergies 109, 179, 261, 262, 268, 270, 285, 297, 324, 326, 336
ALS 57, 58, 114, 261, 290, 296, 404
Alzheimer's 16, 52, 57, 114, 273, 290, 302, 416
Anderson, Chris 131
angina 2, 54. *See also* cardiac symptoms, conditions and recommendations
antioxidant nutritional supplements 189, 307
Arizona Center for Integrative Medicine 10
Arnetz, Dr. Bengt 282, 406
arrhythmia 2, 54. *See also* cardiac symptoms, conditions and recommendations
Asperger's Syndrome 57, 237, 293
asthma 257, 262, 268, 269, 270, 404
autism 4, 11, 14, 18, 57, 58, 106, 114, 175, 237, 273, 291, 292, 293, 295, 296, 297, 298, 299, 300, 301, 302, 304, 305, 308, 324, 325, 367, 387, 388, 391
Autism Spectrum Disorders 57
autism resources 387
EMR, mercury and autism 301
NMT & Autism 324

B

baby monitors 206
Baby Monitors - Safer Solutions 207
baby sensor pads/bedding 207
Baby Sensor Pads/Bedding - Safer Solutions 208
baby's room 205
Baerlocher, Mark (MD) 192
Balch, James (MD) 193, 400
Balch, Phyllis A. (CNC) 193, 400
Balmori, Alfonso (PhD) 93
Bateman, Robert 348
Becker, Robert (MD) 12, 17, 76, 187, 318, 359, 400
bed canopies 232
bio-electrical field 106
BioInitiative Working Group/Report 44, 101, 264
Biological Dentistry 310
  Biological Dentists 391
biological effects 39, 50, 64, 65, 153
birds' ability to migrate 91
birth defects 288
Blackman, Dr. Carl 52, 405
Blackwell, Grahame (PhD) 7
Blank, Martin (PhD) 16, 25, 31, 46, 50, 78, 79, 176, 177, 252, 412
blood-brain barrier 55, 64, 65, 69, 222, 259, 303, 307, 324, 403, 408, 411, 416, 420
Body Electric 12
Boham, Elizabeth (MD) 388
books 400
brain cells 59
brain tumors 52
breast cancer 59

C

calming the scattered mind 332
Canadian Initiative to Stop Wireless, Electric, and Electromagnetic Pollution (WEEP) 102
cancer 16, 17, 20, 33, 36, 45, 49, 58, 59, 61, 62, 63, 70, 74, 86, 87, 111, 167, 176, 183, 189, 190, 191, 212, 213, 248, 252, 260, 261, 284, 285, 305, 360, 362, 366, 389, 404, 405, 407, 408, 411, 412
cardiac symptoms, conditions and recommendations 2, 154, 262, 286, 287, 319, 404
Carlo, George (PhD) 34, 35, 72, 90, 177, 300, 323
Carpenter, Dr. David 86
Carson, Rachel 12, 18, 91
cataracts 262, 288, 404
cathode ray terminal 134
cathode ray tubes (CRT) 196
CDC (Centers for Disease Control and Prevention) 58
cell membrane 305
cell phone contract 48
cell phone headsets 161
cell phones, PDAs 147
   pregnant women, children and cell phones 149
cell tower antenna 22
Chekhov, Anton 92
chemical hazards 17
chemically-scented candles 18
chemical, or other environmental, sensitivities 261
chemical sensitivity 249
Cherry, Dr. Neil 51, 386, 412
Chevalier, Dr. Gaétan 318
children and cordless phones 144
Children & Nature 387
chips, pendants 229
chromosome damage 92
Chronic Fatigue Syndrome 183, 274–277
circadian rhythms 178
citizens' groups 104
Colony Collapse Disorder 90. *See* disappearance of bees
Colwood, British Columbia 97
Communication Challenge 338
compact fluorescents 23
complementary medicine 314
Computer Game Consoles 167
computer screens 196
concentration problems 52
cordless phone 132, 138, 251
   6.0 GHz cordless phone 132
   900 MHz cordless phone 146
   Children and Cordless Phones 144
   Cordless Phone Models: 142
   Cordless Phones - Safer Solutions 145
   digital cordless phone 133
crib death 208
crib toys 209
Cross Currents - The Perils of Electropollution, The Promise of Electromedicine 12
CRT monitors 196
CT (computed tomography) scans 191
   Medical Scans - Safer Solutions 192

D

Dacre, Sarah (ES-UK) 102, 112, 258, 259, 265, 383
dairy cows 92
Danish Study 40
Davis, Professor Devra (PhD) 25, 46, 63, 71, 378, 379, 400
Dean, Carolyn (MD, ND) 25, 275, 276, 277, 305, 306, 400
DECT 138
demand switch 232
dementia 59
depression 52, 178, 262, 273, 281, 283, 326, 404
detoxification 305
digital TV 201, 202
digital X-ray 189
dirty electricity 23, 183
disappearance of bees 90
dizziness 33, 52, 57, 109, 113, 179, 237, 259, 262, 315, 404, 408, 409
dizzy spells 234
DNA 33, 53, 189
   damaged DNA 53
Dowson, Dr. David 22
Doyon, Raymond Paul (MA) 37, 260
dryers 250
DVD and video players and recorders 202

E

EEA 44, 77
EEG 52
EHC-D protocol 264
EHS. *See* electro-sensitivity
Einstein, Albert 90
electrical - wiring and fuse boxes 213
  Fuse Boxes - Safer Solution 214
electric equipment 212
electric field meters 395
electric heating pads, electric blankets 203
  Electric Blankets, etc. - Safer Solutions 204
electric hybrid car 23
electric in-floor heating 245
electric razors 209
  Electric Razors - Safer Solutions 211
Electromagnetic Health.Org 102
electro-pollution 106
Electro-sensitivity and CFS 278
electro-sensitivity (ES)/electro-hypersensitivity (EHS) 14, 33, 102, 109, 178, 205, 257, 259, 262, 275, 278, 295, 308, 314, 367, 404
EMF/EMR Consultants 393
EMF Safety Store 397
EMR Association of Australia 104
EMR Policy Institute 101, 381
EMR Testing Equipment 240
energy medicine 318
entertainment devices 201
  televisions 201
Environmental and Resource Studies Program, Trent University 37
environmental influences 59
Environmental Working Group 10, 101, 382
epilepsy 178
Essential fatty acids (EFAs) 306
Ethernet 180
Ethernet connection 105
excessive urination 326
exercise machines 212

F

Facebook 342
Family Action Plan 15
far field 147

Federal Communications Commission (FCC) 46
Feinberg, Leslie S. (DC) 25, 322, 323, 324, 325, 388
fiber optics 88, 365
Finnish Radiation and Nuclear Safety Authority 44
Firstenberg, Arthur 91
fluorescent lights
  Fluorescent Lights Safer Solutions 184
food allergies 268
Food and Drug Administration (FDA) 46
fragrances 268
free radicals 53, 188
French government 44
frequency detection 395
fungal diseases 326

G

gaming 24, 68
  Safer Gaming Solutions 168
Gandhi, Professor Om P. 71, 72, 378
gasoline-powered car 217
  Non-hybrid Cars - Safer Solutions 219
Gee, David 45, 362
Generation Rescue 298, 302
German Federal Agency For Radiation Protection 138
Global Positioning Systems (GPS) 219
Goodman, Dr. Reba 50
Good Morning America 341
Gothenburg 104
Green Vaccines 302
grounding exercise 318
guidelines on RF exposures 45
Guinea Pigs "R" Us 102
Gustavs, Katharina 25, 82, 83, 131
Gust, Larry 25, 131, 135, 162, 206, 268, 272, 299, 393, 403

H

H1N1 178
hair dryers 209
  Hair Dryers - Safer Solutions 211
Haley, Boyd (PhD) 301, 302, 308
Hardell, Professor Lennart 70, 71, 111, 112
Harkin, Senator Tom 95

Hartman, Stan 25, 131, 134, 157, 215, 231, 394
Haumann, Thomas (PhD) 65
Havas, Magda (PhD) 25, 37, 88, 102, 177, 183, 258, 375, 381, 401, 408
headaches 33, 57, 67, 109, 115, 171, 179, 183, 184, 199, 222, 237, 259, 262, 264, 315, 324, 326, 408, 417, 419
headsets 161
   cell phone headsets 161
   Headsets - Safer Solutions 163
   wireless headsets 161
Health Action Network (HANS) 102
Health Canada 70
Healthy Type A - Good News For Go-Getters 9
heart. *See* cardiac symptoms, conditions and recommendations
heartburn and other stomach and digestive problems 326
heavy metal toxicity 302
Herberman, Dr. Ronald B. 86, 149
HESE (human ecological social economical project) 103
high voltage power lines 21
Hoch, Christine 101
Hope For The Future 15
hormones 52
hotel/motel accommodations 227
hot tubs 212
household appliances 204
hybrid cars 216
   Hybrid Cars - Safer Solutions 217
Hydroxy L-tryptophan (5-HTP) 283
Hyman, Dr. Carrie (LAc, OMD) 25, 38, 53, 149, 321, 377, 389
Hyman, Mark (MD) 273, 388, 400
hypertension 2, 54

I

IBE Certified Consultants 392
ICEMS. *See* International Commission for Electromagnetic Safety
ICNIRP: International Commission on Non-Ionizing Radiation Protection 31, 44, 45
   ICNIRP regulations 78
immune system 51, 53, 74, 178, 227, 256, 257, 260, 266, 267, 274, 277, 282, 307, 310, 311, 312, 322, 323, 336, 406, 407
Immune System Impairment 267
infertility 67, 276, 288
inflamed gums 326
in-flight video screens 223
insomnia 4, 237, 260, 262, 264, 276, 279, 280, 283, 285, 321, 326, 404, 408
Institute for Health and the Environment at the University of Albany 86
International Commission for Electromagnetic Safety 101, 386
Internet access 134
   Safer Internet Access Solutions 170
   wireless Internet access 169
Internet Safety 342
Interphone Study 40
irrelevant safety standards 31
irritability 326
itching 326

J

Johansson, Olle (PhD) 1, 26, 28, 31, 58, 73, 74, 112, 177, 290, 364, 378, 413

K

Kabat-Zinn. Dr. Jon 332, 400
Karlsson, Anne-Li 104
Karolinska Institute 28, 282
Kartzinel, Dr. Jerry (MD) 303, 401
Kelley, Elizabeth (MA) 100
Kelsey, Dr. Frances 43
Kenton, William 91
Key to the Safer Solutions 134
Khurana, Vini (MD) 26, 66
Klinghardt, Dietrich (MD) 26, 296, 297, 302, 306, 308, 325, 389

L

Lai, Dr. Henry 38, 204, 210, 420
landline phone 145, 163
laptops 180, 197
Last Child in the Woods 199
LCD screen 134, 197
learning and memory 178
Lee, Dr. Lita 137
leukemia 60

childhood leukemia 60
library mum 233
Liechtenstein Parliament 100
Lipsitz, Dr. Jeffrey 282
Litovitz, Dr. Ted 51
Louv, Richard 199
low EMR environment 227
Low Tech Family Challenge 14
Low-Tech Family Challenge: 344, 351
Lucas, Caroline (MEP) 97
Lyme Disease 275

M

magnesium 276
Martin, Dr. Keith 97
Martínez, Alfonso 91
mast(s) 22
Mast Sanity 103
May, Elizabeth 97
McCarthy, Jenny 175, 302, 401
McGill University 61
McKinney, Dr. Heather 20, 26, 112, 373, 376
medical clinics/professional practices 388
medical X-rays 188
melatonin 307
memory loss 52
mercury 18
  mercury amalgam dental fillings 302
Metzinger, Rob 26, 131, 137, 148, 238, 253, 392
Microwave News 40
microwave ovens 135
  Microwave Ovens - Safer Solutions: 137
  microwaving baby milk 137
Microwave-Syndrome 222, 315
microwave transmitter boxes 202
Miller, Jesse 342
Moms Against Power Lines 215
Moms for Safe Wireless 101
Morgan, Lloyd 42, 402
Mosgoeller, Wilhelm (MD) 64
MRI scans 191
  Medical Scans - Safer Solutions 192
MS (Multiple Sclerosis) 57, 261, 276, 290, 389, 404
multiple chemical sensitivity. *See* chemical sensitivity

myalgic encephalomyelitis 274.
  See Chronic Fatigue Syndrome
My Wireless Wake-up Call 16

N

natural brands of cleaning products 399
nausea 52
near field 147
networking your Ethernet 171
network settings 170
NeuroModulation Technique, NMT 322
  NMT & Autism 324
neuronal (brain) damage 56
Newton, Janet 101
non-thermal levels 33

O

Oberfeld, Dr. Gerd 100, 177, 222, 361, 417
Öberg, Jennie 104
O'Connor, Eileen 101, 102, 360
office equipment 200
  Office Equipment - Safer Solutions 200
  photocopiers 200
  printers 200
Omega-3 and -6 fatty acids 306
opthalmological/vision problems 288

P

pagers 157
Paracelsus Clinic 260
Parkinson's 57, 58, 115, 260, 261, 290, 296, 416
pH Balance 326
Philips, Alasdair 26, 102, 131, 141, 150, 169, 221, 223, 244, 375
photocopiers 200
physical radiation barriers 231
plasma screens 201
portable music players 203
Porto Alegre Resolution 97
power cords 215
  shielded power cord 232
power lines 215
  Power Lines - Safer Solutions 215
power tools 213
Powerwatch 102, 141, 169, 288, 375
precautionary principle 97

pregnancy 61
pregnant women, children and cell phones 149
premature ageing in pine needles 92
pre-natal ultrasound 187
   Ultrasound Safer Solution: 188
President Obama 360
President's Cancer Panel 101
printers 200
   ink-jet printers 200
Prior, Senator Mark 95
probiotics 275
protective shielding 396

R

radiation detector 22
radiation from mobile phones 222
Radiation Rescue Communications Kit 166
Radiation Rescue Early Warning Cue 317
Radiation Rescue Intentions 236
Radiation Rescue Questionnaire 111
Radiation Rescue Retreats 227
Radiation Rescue - The Essentials 145
Radiation Research Trust 101, 360
Radio Frequency Guidelines. *See* RF Guidelines
Radio Frequency ID tags 193
   ID Tags - Safer Solutions 194
radio frequency radiation (RF) 79, 224
   guidelines on RF exposures 45
Rau, Thomas M. (MD) 260, 275, 308, 309, 389
Rea, William (MD) 26, 187, 253, 264, 267, 374, 389, 401
red blotches 326
Rees, Camilla 102, 381, 401
Rees, Dr. Allison 74, 338, 339, 340
Reference Material 402
Regulations and Safe Levels Dispute 32
Reviewing Our Radiation Rescue Intentions 355
RF Guidelines 80
RF (radio frequency) meter 240, 395
Room by Room - Audit & Recommendations 238
room deodorizers 18

S

Safe Levels 78
Safe Living Technologies 396
Safer Internet Access Solutions 170
Salford, Leif (MD) 29, 55, 56, 290, 324, 414, 417
Salti, Roberto 344
Sanoviv 309, 390
SAR - Specific Absorption Rate 152
Sasco, Dr. Annie 37
Saunders, Mayor David 97
Scheiner, Ana 390, 401
Scheiner, Hans (MD) 26, 69, 109, 141, 165, 221, 222, 252, 259, 260, 261, 263, 264, 281, 305, 306, 307, 314, 315, 316, 324, 390, 401, 416, 419
Schlegel, Peter 80, 82
security 226
security full-body scans 195
   Security Scans - Safer Solutions 195
Sensory Perspective 230, 398
Servan-Schreiber, Dr. David 37
sex ratio 290
shielded power cord 232
shielding materials 231
   shielding curtains 232
   shielding fabrics, wallpapers, paints 232
SIDS 208
Sierck, Peter 27, 84, 392
Silent Spring 12, 18
Sinatra, Stephen (MD) 12, 17, 27, 52, 54, 154, 287, 318, 319, 390, 401
skin inflammation 326
sleep 279
sleep disorders 205, 418. *See also* insomnia
sleep zone 242
Slesin, Louis (PhD) 27, 40, 44, 46, 63, 377, 382
SmartGrids 88
Snively, John (DDS) 310
Specter, Senator Arlen 95
Sprott, Dr Jim (OBE) 208
Starkey, Sarah (PhD) 27, 178, 181, 376, 384, 413
Statement of Accountability 362
Step 1 Know The Evidence 13
Step 2 Know Your Risks 14
Step 3 Seek Safer Solutions 14

Step 4 Enjoy Your Family Action Plan 14
stove 249
stress disorders 281
Stress Reduction Clinic at the University of Massachusetts Medical Centre 332
sunbeds 212
sunlamps 212
supplements 307

T

tachycardia 2, 54, 259. *See also* cardiac symptoms, conditions and recommendations
Teitelbaum, Jacob (MD) 27, 277, 278, 283, 284, 301, 305, 306, 327, 391, 401
televisions 201
TENS 213
testicles 288
testing and shielding equipment 395
testing of your home, office and school 238
text messaging 150
Thermal and Non-Thermal 30
thermal levels 31
Thimerosal 302
Tourtet, Christiane 96
Traditional Chinese Medicine (TCM) 320
TV and video remote controls 202
Type A 330

U

ultrasound. *See* pre-natal ultrasound
US Federal Regulations 46

V

vaccines 18, 178, 257, 273, 292, 302, 310
Vacuum Cleaners 250
vehicles 216
video games 68
Video/Wireless Games 167
  Safer Gaming Solutions 168
viruses 178
Volkrodt, Wolfgang 92

W

Warnke, Dr. Ulrich 94
Warren, Diana 101
washing machines 250
Weatherall, Martin 102
Weil, Dr. Andrew (MD) 10, 204, 314
Wentz, Dr. Myron 309, 390
wildfires 92
wildlife 93
WiredChild 7, 102, 384
wired devices 196
wireless communication 59
wireless earpiece phones 159
wireless free accommodations 227
wireless headsets 161
wireless in aircraft 24, 221, 416
wireless Internet access 169
wireless radiation 172
wireless radiation and reaction time 222
wireless router 23, 251
wireless systems in schools 175
wLAN 180
World Health Organisation 37, 44, 70

X

X-rays
  dental X-rays 189, 190
  medical X-rays 188
  Medical X-rays - Safer Solutions: 190

Y

yeast 275

Z

Zaret, Dr. Milton 288
Zhu, Dr. Hong Zhen 320, 321, 401